T0290614

Interpreting Energy at Museums and Historic Sites

AASLH

INTERPRETING HISTORY

SERIES EDITOR

Rebekah Beaulieu, Taft Museum of Art

MANAGING EDITOR

Aja Bain, AASLH

EDITORIAL BOARD

William Bomar, University of Alabama Museums
Jessica Dorman, The Historic New Orleans Collection
Harry Klinkhamer, Venice Museum & Archives
Anne Lindsay, California State University–Sacramento

Steven Lubar, Brown University
Laura A. Macaluso, York County History Center
Ann McCleary, University of West Georgia
Porchia Moore, Johns Hopkins University
Debra Reid, The Henry Ford
Laura Roberts, Roberts Consulting
Kelby Rose, Independent
Zachary Stocks, Oregon Black Pioneers

William Stoutamire, University of Nebraska–Kearney

William S. Walker, Cooperstown Graduate Program SUNY Oneonta

ABOUT THE ORGANIZATION

The American Association for State and Local History (AASLH) is a national history membership association headquartered in Nashville, Tennessee, that provides leadership and support for its members who preserve and interpret state and local history in order to make the past more meaningful to all people. AASLH members are leaders in preserving, researching, and interpreting traces of the American past to connect the people, thoughts, and events of yesterday with the creative memories and abiding concerns of people, communities, and our nation today. In addition to sponsorship of this book series, AASLH publishes *History News* magazine, a newsletter, technical leaflets and reports, and other materials; confers prizes and awards in recognition of outstanding achievement in the field; supports a broad education program and other activities designed to help members work more effectively; and advocates on behalf of the discipline of history. To join AASLH, go to www.aaslh.org or contact Membership Services, AASLH, 2021 21st Ave. South, Suite 320, Nashville, TN 37212.

ABOUT THE SERIES

The American Association for State and Local History publishes the *Interpreting History* series in order to provide expert, in-depth guidance in interpretation for history professionals at museums and historic sites. The books are intended to help practitioners expand their interpretation to be more inclusive of the range of American history.

Books in this series help readers:

- quickly learn about the questions surrounding a specific topic,
- introduce them to the challenges of interpreting this part of history, and
- highlight best practice examples of how interpretation has been done by different organizations.

They enable institutions to place their interpretative efforts into a larger context, despite each having a specific and often localized mission. These books serve as quick references to practical considerations, further research, and historical information.

TITLES IN THE SERIES

Interpreting Energy at Museums and Historic Sites

Leah S. Glaser

ROWMAN & LITTLEFIELD
Lanham • Boulder • New York • London

Published by Rowman & Littlefield
An imprint of The Rowman & Littlefield Publishing Group, Inc.
4501 Forbes Boulevard, Suite 200, Lanham, Maryland 20706
www.rowman.com

86-90 Paul Street, London EC2A 4NE

Copyright © 2023 by The Rowman & Littlefield Publishing Group, Inc.

All rights reserved. No part of this book may be reproduced in any form or by any electronic
or mechanical means, including information storage and retrieval systems, without written
permission from the publisher, except by a reviewer who may quote passages in a review.

British Library Cataloguing in Publication Information Available

Library of Congress Cataloging-in-Publication Data
Names: Glaser, Leah S., author.
Title: Interpreting energy at museums and historic sites / Leah S. Glaser.
Description: Lanham : Rowman & Littlefield Publishers, 2023. | Series:
   Interpreting history | Includes bibliographical references and index.
Identifiers: LCCN 2022042262 (print) | LCCN 2022042263 (ebook) | ISBN
   9781538150535 (cloth) | ISBN 9781538150542 (paperback) | ISBN
   9781538150559 (epub)
Subjects: LCSH: Power resources—United States—Exhibitions. | Electric
   power production—United States—Exhibitions. | Industrial
   museums—United States. | Museums—Educational aspects—United States. |
   Historic sites—Interpretive programs—United States.
Classification: LCC TJ163.155.A1 G537 2023 (print) | LCC TJ163.155.A1
   (ebook) | DDC 621.0420973—dc23/eng/20221012
LC record available at https://lccn.loc.gov/2022042262
LC ebook record available at https://lccn.loc.gov/2022042263

∞™ The paper used in this publication meets the minimum requirements of American National
Standard for Information Sciences—Permanence of Paper for Printed Library Materials, ANSI/
NISO Z39.48-1992.

Contents

Acknowledgments

I have been thinking about the contents of this book for some time. Before I began working at the water and power utility, the Salt River Project (SRP) in Phoenix, while in graduate school at Arizona State University in 1995, I certainly never considered the possibility that I would be writing about energy. I took the internship to focus on accessing sources that told the story of the Yaqui, part of a tribe from Mexico who had settled near Phoenix and worked for SRP, for my MA thesis. But my supervisor, Dr. Karen Smith, had the confidence in me to handle a HAER report for a transmission line, despite my having no engineering background. The deep dive into long-distance transmission towers put me on my research path into the history of electrical power, particularly issues of distribution and equity. I began writing my dissertation on rural electrification (*Electrifying the Rural American West: Stories of Power, People, and Place*) amidst the Enron scandal and California's rolling black-outs. Only a handful of historians had begun to address the topic, and several far more impressive scholars have since added to the historiography of electricity and energy in the last two decades. This book is an attempt to introduce the public to that research and scholarly interpretation beyond books, podcasts, and documentaries.

So, I am not a physicist. This work aims for basic historic understanding of energy, and I have done my best. Since graduate school, I have been writing and researching topics both directly and tangentially related to energy. While my research and publications initially focused more on the technological and sociocultural aspects, my work in public history and historic preservation showed me the many ways various aspects of public history and cultural resource stewardship could inform issues about environmental sustainability. Writing this book has allowed me to collate and refocus dozens of Word files at various states of publication over the last quarter century, including reports, lecture notes, journal notes, conference papers, reviews, and actual publications.

This volume is not quite what I first envisioned and may only have some features in common with others in the series. Limited ability to travel and visit historic sites and museums, mostly due to the COVID pandemic, geographically limited my range of case studies. That I was able to find so many examples within a drivable range, however, is encouraging. However, "lockdown" conditions at least allowed me the ability to connect to several generous energy scholars and public historians across the country who provided knowledge, references, opinions, impressions, advice, and references. Therefore, it is by no means comprehensive, but rather suggestive, offering context and some ideas and perspectives for interpreting energy not just scientifically, but a bit more historically. At the same time, there is so much more to say about the topic, especially on a global scale. This is a small first step. Please build on it.

I have read and written much about the history of energy in my career, but I am hardly an expert. I want to thank the American Association of State and Local History's reviewers, as well as Aja Bain, for her support and copyediting; Rowman and Littlefield's Charles Harmon, who kept things moving; and AASLH's Debra Reid and David Vail, for their enthusiasm, encouragement, and whose own contributions to this series inspired me to write this one. I am so incredibly grateful to all these scholars who generously shared their time, knowledge, impressions, and expertise: Robert Bauman, Carolyn M. Crawford, Evan Finch, Robert Franklin, Devin Hunter, Christopher Jones, Andy Kirk, Jason Krupar, Jeff Manuel, Linda Richards, and Joan Zenzen. Any inaccuracies reflect my own misunderstandings. Other site and resource managers—including Carrie Barske with Muscle Shoals, Chris Gasiorek of the Mystic Seaport Museum, Kris Kirby from the Manhattan Project National Historic Site, Rob Verbsky from the Owl Head Transportation Museum, Andrew Rowand at the Kent Iron Furnace, and William Brown, Ryan Paxton, and Andrew Sargent of the Eli Whitney Museum—provided on-site expertise. The National Association of Interpretation's Karin Hostetter provided a fabulous foundation in best practices. Huge thanks to Chris Magoc, Samantha Boardman, and Steve Burg for granting me permission to reprint their previous work. Others, like Historian of Science Linda Richards and the Salt River Project's Leah Harrison, spent the time writing pieces especially for this publication. It was my time at the Salt River Project that led me to energy history in the first place. Special thanks to the staff at Legacy Management for thoughtfully answering questions and describing their vision and exhibit for the Atomic Legacy Cabin. I'll add gratitude to my Airbnb hosts, Frank and Michaelanne, in Rockland, Maine, for providing me with a lovely environment to kick off the difficult task of writing. And for helping me to secure last-minute images: Jaime Aguila; Christina Volpe; and Molly Norton (The Barnes Museum) and Jodi Silvio at the Salt River Project.

Lastly, of course, I must thank my precious family: my husband and fellow historian, Steve Amerman, for his consistent support and patience; my daughter, Meredith, whose interest in the environment inspired me to try to do what I could about preserving it; and my son, Ben, who, while tired of visiting museums and historic sites, shares an affinity for architecture, engineering, and design. You all inspired so many ideas, even if it wasn't intentional.

Preface

Why Interpret Energy History?

In 2018, teenage climate change activist Greta Thunberg began pleading with adult audiences throughout the world to consider the kind of planet we are leaving to our children. Likewise, ensuring that resources remain accessible to those same future generations is one of the most important reasons for preserving and interpreting historic sites. Outside of political contexts, historic exhibits and places can highlight particular narratives that stress change over time, in service to how we make future narratives about energy use. Stewards of heritage sites and collections can engage the public at the grassroots level to raise awareness about the cultural and socioeconomic reasons for past choices that have contributed to climate change. In a 2021 interview at the Edinburgh TV Festival, Thunberg observed that "we cannot overestimate the power of storytelling" in changing social norms.[1]

Climate change has emerged as one of the chief concerns for today's heritage organizations across the globe. Pollution and acid rain have eroded historic structures throughout the last century, but the increased frequency of violent and destructive storms regularly threatens the natural and cultural resources along shorelines from Louisiana to New Jersey to Scotland to China. Drought and subsequent wildfires grow increasingly frequent, also destroying historic and cultural sites. Scientists have even designated a new geological period, the Anthropocene, to identify an era in which human beings have significantly altered the climate, primarily through the heavy use of nonrenewable and carbon-producing energy fuels and systems. In May 2022, scientists measured carbon dioxide levels at Hawaii's observatory at over 50 percent higher than pre-industrial levels.[2] We know now that we cannot reverse global warming, but we still must slow climate change and adjust to how we live today and in the future.

Yet, as extreme weather events grow more frequent and severe, so has political resistance to addressing our energy supply. The scientists' warnings that fossil fuel use has changed, and continues to change, the planet's climate and ecology contradict the historic values, cultural practices, and powerful historical narratives that have indoctrinated so many of us, including policymakers, for so long. Energy conservation is politically unpopular. Between 2017 and 2021, the American government rolled back over a hundred environmental policies and laws.[3] In 2022, the Supreme Court ruling in *West Virginia vs. Environmental Protection Agency* limited the ability of the government to regulate emissions, and Congress struggled to pass legislation to address climate change. Modern society uses energy unsustainably and depends largely on nonrenewable sources that have shaped our built environment, our infrastructure, our material culture, our values, and how we live. As author Michael Webber observes, "Energy is unique: no other physical factor in society has such a wide-ranging impact on public health, ecosystems, the global economy, and personal liberties."[4] This has been especially true over the last century. Back in 2001, amidst widespread blackouts in California, Republican president George W. Bush's presidential spokesman Ari Fleischer rejected a reporter's suggestion that perhaps Americans should curb their energy use:

> That's a big no. The President believes that it's an American way of life, and that it should be the goal of a policy maker to protect the American way of life. The American way of life is a blessed one. And we have a bounty of resources in this country.

He continued with a nod to energy conservation, but as a choice dictated by the market economy:

> What we need to do is make certain that we're able to get those resources in an efficient way, in a way that also emphasizes protecting environment and conservation, into the hands of consumers, so that they can make choices.

He also tied energy use and its production to American identity:

> But the President also believes that the American people's use of energy is a reflection of the strength of our economy, of the way of life that the American people have come to enjoy. And he wants to make certain that a national energy policy is comprehensive, includes conservation, includes a way of life that has made the United States such a leading nation in the world.[5]

Energy has powered the global economy and dictates daily life, but most of us do not see it or understand it. We just use it—a lot of it. Too much, in fact. Humans have consumed more energy in the last 125 years than in any time prior (when we relied on human and animal muscle, wood, and water). Energy fuels have become almost as essential as food and water to daily human survival. Access to energy infrastructure largely determines economic equality, social equity, environmental justice, or most critically, national security.[6] When Russia, one of the world's largest exporters of gas and oil, invaded Ukraine, a major producer of nuclear power, in February 2022, the action sparked global outrage and fear. Countries issued sanctions, creating an energy crisis with skyrocketing fuel costs. It is difficult to over-state the importance of energy in the world economy, geopolitics, and daily modern life. Entire regional communities and cultural practices have developed around the extraction and the processing of energy sources, and there remains a deep emotional and cultural attachment to those lifeways. We therefore cannot solve climate change with only top-down policies and goals. We also need to make changes from the bottom up, and when we make those changes, we need to understand how energy choices impact different communities.

Arguing that "the hegemony of large systems is culturally shaped" and that our high-energy lifestyle resulted from "a confluence of cultural choices," historian of electricity David Nye concluded that "access to energy had become an inalienable American right, for the individual had been literally empowered by new power sources."[7] Energy production is engrained in colonial Puritan and antebellum values of industriousness. Subsequently, modes of energy became the byproducts of the Progressive Era's efficient use of natural resources "for the greater good." These narratives inform the choices that we as a society make about energy use—including what kind and how much—and impact our willingness to embrace doctrines of environmental sustainability, to strive toward environmental justice, and to engage in large-scale energy conservation and transition.

Beyond the borders of historical heritage, the late twentieth-century environmental movement highlighted the dangers of industrial-era energy systems to the natural and cultural environments (marked by the observance of the first national Earth Day on April 22, 1970). The public history profession, founded at the same time as the energy crisis of the 1970s, is an obvious partner in educating a culturally resistant public about a needed shift in how energy impacts the natural and built environment. Societies often enlist storytelling to maintain traditions, but as Greta Thunberg points out, storytelling can also inspire change. This book aims to support that effort at sites with both obvious, and not-so-obvious, connections to energy, as well as to inform stewards unfamiliar with its history.

Organization and Themes

For many reasons, staff at historic sites and museums often overlook or avoid energy as a theme in favor of social or technological history. Yet over the last few decades, historians have responded to the climate crisis with research, critical discussions, and case studies that might help shake up the traditional progress narratives of technological determinism. Those who work in public forums can therefore access a wealth of potential stories, context, and content for developing interpretation around important themes, factoids, and quotes to populate tours and programs at our museums and historic sites. Interpreting energy only within a fixed time frame (often a site's period of significance) prevents visitors from connecting energy use to its long-term impact and how we produce and use energy today. In contrast, contextualizing sites and attractions across a broad historical timeline of energy history can reveal that energy production is woven into cultural practice, economic activity, and historical identity.[8]

Upcoming chapters, organized by energy source, provide basic background on the science of energy extraction and production, but primarily offer guidance to interpret energy use and transition over broad timelines within historical context. Each will begin with a review of histories and historiographies about various forms of energy use and development, primarily in the United States. This book is, therefore, by no means a comprehensive review of the history of energy, but I have drawn from a wealth of scholarship on energy history. These particular narratives stress that energy systems are a byproduct of people's ideas about work and environment, reflected in the culturally influenced decisions and behavior of culturally diverse populations. My appraisal of these energy resources, processes of development, and historical contexts is somewhat superficial in light of the rapidly expanding historiography on the topic, but these chapter reviews likely offer more detail than one would convey in an interpretive program. The goal here is to provide professionals with a foundation and some resources to build relevant programs around the topic of energy. These discussions will offer interpretive guideposts by directing readers toward historical themes, context, and examples that highlight connections between sources of energy and their use.

Many of these histories provide economic and social context that "people" these energy processes and tie them to economies, cultures, and places, most often through the following five interpretive energy themes:

1. Generation and Development
2. Domestic and Industrial Use
3. Economic and National Security
4. Access and Equity
5. Transition and Diversification

Situated within these broad contexts and themes, case studies highlight how even historic sites featuring energy resources are still resolving interpretive conundrums, or have reached creative solutions, for interpreting historic energy technology and use. However, readers should recognize that examples and debates on energy interpretation across historic sites is uneven, so the number of case studies and the depth in which I explore them for different energy types varies. Still, these case studies will collectively offer some best practices that I highlight in the book's conclusion.

Next, because museums and historic sites are collections- and place-based, each chapter concludes with an "artifact spotlight," which suggests potential resources and spaces for interpretation. These include tangible places and objects with which visitors might interact. A final annotated bibliography provides references for engaging with the scholarship of energy history, identifying themes, and thinking critically about global energy use, and throughout American history in particular.

By specifically addressing historic sites, attractions, and resources that feature historic energy use and resources, this book joins AASLH's "Interpreting History" series as a companion or supplement to its other publications that have addressed environmental sustainability. Public historians Debra Reid and David Vail produced an illuminating primer (2019) for AASLH on interpreting the environment at various types of historic sites and telling stories in new and compelling ways. While she did not directly address climate change, Julia Rose's *Interpreting Difficult History at Museums and Historic Sites* (2016) explores how to design programs that challenge visitors' expectations with content about uncomfortable topics such as environmental justice. With the American Alliance of Museums, Sarah Sutton and Elizabeth Wylie pioneered a primer and a toolkit for "the green museum" to aid cultural institutions in enlisting sustainable and energy-efficient practices across operations.[9]

Each institution has unique cultural, financial, and administrative challenges. Thus, what follows in this volume is intended as only the *start* of discussions and strategies for not just reflecting on how we use energy personally and institutionally, but also expanding, complicating, and incorporating histories of energy at our museums and historic sites. While there is urgency in addressing climate change, we are far behind in addressing the cultural barriers to changing how we understand energy and energy use. Cultural institutions that already court a large and trusting audience may also find value here to identify some ways that explore and interpret new stories through previously overlooked resources and programming, especially if staff have made the commitment to "go green."

Notes

1. Greta Thunberg with Cal Byrne, *The Independent*, https://www.independent.co.uk/tv/climate/greta-thunberg-climate-crisis-edinburgh-v3ae62387, accessed June 29, 2022.
2. National Oceanic and Atmospheric Administration, U.S. Department of Commerce, "Carbon Dioxide Now More Than 50% Higher Than Pre-industrial Levels," June 3, 2022, https://www.noaa.gov/news-release/carbon-dioxide-now-more-than-50-higher-than-pre-industrial-levels.
3. Nadja Popovich, Livia Albeck-Ripka, and Kendra Pierre-Louis, "The Trump Administration Rolled Back More Than 100 Environmental Rules. Here's the Full List," *New York Times* (January 20, 2021).
4. David Nye, *The American Technological Sublime* (Cambridge: MIT Press, 1996); Michael E. Webber, *Power Trip: The Story of Energy* (New York: Basic Books, 2019), 2.
5. Press briefing by Ari Fleischer, the James F. Brady Briefing Room, Office of the Press Secretary, The White House of President George W. Bush, May 7, 2001, https://georgewbush-whitehouse.archives.gov/news/briefings/20010507.html, accessed February 24, 2020.
6. David E. Nye, *Consuming Power: A Social History of American Energies* (MIT Press, 1998); Webber, *Power Trip*, 6–8.
7. Nye, *Consuming Power*, 5, 249–51.
8. Paul Sabin, "'The Ultimate Environmental Dilemma': Making a Place for Historians in the Climate Change and Energy Debates," *Environmental History*, Vol. 15, No. 1 (January 2010), 76–93.
9. See Sarah Sutton (Brophy) and Elizabeth Wylie, *The Green Museum: A Primer in Environmental Practice*, second edition (Lanham, MD: Rowman and Littlefield, 2013). "The Climate ToolKit" is a collaborative of cultural institutions committed to "aggressively address climate change" within the categories of energy, food service, transportation, plastics, landscapes and horticulture, investments, visitors, and research. The efforts support and align with the United Nations Sustainable Development Goals. Climate ToolKit, https://climatetoolkit.org, accessed December 17, 2021.

CHAPTER 1

Introduction

Interpreting Energy History

IN HER BOOK *Environmental Sustainability at Historic Sites and Museums,* Sarah Sutton discusses how historic sites and museums can make sustainable choices about their own energy systems, both within and outside of interpretive activities. Using examples such as biothermal energy systems at the Hanford Mills Museum in the Catskill Mountains of New York, Sutton stresses that energy interpretation can refer to present systems at a site as well as those in the past. She further provides examples for how interpretive programs can exercise reflexive practices by being mindful of energy efficiency, whether enlisting a wood-burning fire or engaging in a canning process.[1]

Sutton's work meets the needs of cultural heritage organizations that have sought ways to reduce carbon emissions and adopt environmentally sustainable practices and values into their missions over the past couple of decades.[2] The National Trust for Historic Preservation, the George Wright Society, and the National Park Service (NPS) have issued several reports on climate change. Even the Union of Concerned Scientists issued a threatened cultural properties list as early as 2014.[3] Agencies like the NPS and the National Association for Interpretation (NAI) are holding professional development workshops about interpreting climate change. Alongside the New Building Institute and the New England Museum Association, the nonprofit Environment and Culture Partners have also urged cultural institutions to lead by example through an initiative known as "Culture over Carbon."[4]

Linking institutional priorities of cultural and environmental stewardship to compelling historical interpretation requires us to confront many past associations and cultural assumptions, not to mention a fraught and divided political landscape. The traditional, popular narrative celebrates technology and its machines as essential tools of social and economic progress that use and extract energy. School textbooks and popular culture have historically focused upon and celebrated narratives chronicling how technological progress begets economic prosperity and cultural advancement. The images of converting raw materials to energy for movement, heat, and light are iconic symbols of this progress across American landscapes. David Nye coined the phrase "the American technological sublime" to describe those iconic symbols that are often integral to sense of place and regional identities. They include the rotating water wheel beside the mill, the gently smoking iron furnace in the forest, the steamboat on the river, the railroad train across the mountains, the windmill on the farm, and even power lines across empty landscapes and highways. Many sites that interpret resources like these have often reinforced visitors' nostalgic perceptions of energy and complacency about it.

What we choose to remember, interpret, and preserve reflects both our society's evolving cultural values, as well as our resistance to change. The preservation movement in the United States began with private citizens saving, rebuilding, and developing Mount Vernon, Williamsburg in Virginia, and other colonial homes as historic sites against the rapid changes of industrialism. Industrial museums such as the Baltimore Museum of Industry in Maryland, the Museum of History and Science in Seattle, Washington, and the Lowell National Historic Site in Massachusetts have grown more popular with deindustrialization and suburbanization. Nostalgic feelings about community, home ownership, and job security focused interpretive content on technological progress. The public likewise tends to associate energy-powered industry and technological innovation more with economic prosperity than the impact on natural resources.

Some stewards of historic places, resources, and machinery have appealed to policy makers, developers, and planners to recognize historic preservation's key role in supporting and advancing the values of sustainability with arguments on energy savings, material recycling, and high-density urban planning.[5] This argument and the tagline "The Greenest Building is the One Already Built" has supplemented, and in some cases replaced, the economic argument for preservation based on heritage tourism, adaptive reuse, and property values.[6] Planners, administrators, and preservationists can enlist this volume for identifying, documenting, contextualizing, and preserving historic resources about energy generation, use, and environmental impact amidst the rapidly changing climate emergency. Many stewards of historic technologies have already addressed environmental concerns. The managers of the historic Collins Axe powerhouse in Connecticut, for example, replaced its turbines to avoid burning fossil fuels.[7] But in the area of *interpretation*, many historic sites have been slow to translate activism among colleagues into challenging public audiences to consider sustainability, energy conservation, and transitioning to new energy sources and systems. Emphasizing the dynamic nature of both energy technologies and energy use can reveal historical foundations from which we can learn to adapt to new energy systems.

To complicate matters, both the interpretation and the preservation of historical energy resources sometimes conflict with today's efforts at environmental conservation. Sites might feel compelled to interpret a process that was environmentally destructive in the past, while living in an age that necessitates us to think in terms of environmental sustainability. In his book about air quality, environmental historian David Stradling ruminated about "how to use the troublesome reality of smoke to destroy the image of smoke as a sign of progress without threatening the idea of progress itself."[8] Steamboats, trains, mills, iron furnaces, factories, and powerplants are environmentally problematic historical resources to interpret and preserve. But as historically significant resources that transformed how we live, these landscapes and machines also tell stories about historic energy, energy processes, energy systems, and most importantly, our use of natural resources for energy in exploitative ways. While many historic sites have celebrated or commemorated the technology of machines, they do not often reflect upon or engage their audience in contemporary discussions about those technologies. This is not surprising, since climate change is highly politicized. However, interpretation around the theme of energy can address issues highly relevant to climate change in a way that engages the public about past behavior and choices, rather than overtly challenging present beliefs.

How to accomplish this poses lots of questions. How can museums and historic sites communicate those stories, while at the same time modeling and advocating for just a little more cognizance about conserving energy? How might they support informed decisions about transitioning away from fossil fuels? How might historic sites highlight models of adaptability and resilience while also providing nostalgic, enjoyable, and inspiring visitor experiences? How do curators preserve historic energy resources when we know that burning fossil fuels causes climate change? How should frontline interpreters balance the aversion to perceived climate activism at museums (as "objective" historians) with their responsibility to convey facts? How do we inspire visitors to adopt a sense of stewardship not only for our historic cultural resources, but for the whole earth, as well?

By enlisting traditional historical inquiry and research alongside interpretation and collaboration, public history often tries to address and inform current and future problems. Effectively designed exhibits and programs can provoke public audiences to think critically about past values in a way that can influence future policy. Many, but certainly not all, historic sites and museums might be able to enlist artifacts from their collections to interpret energy. On October 30, 2020, *The New York Times* printed a story that assessed each of the states across America with their primary source of electricity. The article conveys that energy resources are extremely diverse across not just time, but also place. In other words, energy development and use are both time-specific and often place-based.[9]

To stay relevant and sustainable as a profession, public historians should continue to address what attracts and excites public audiences, ideally by connecting to the place-based experiences and knowledge that individual visitors or groups bring to a site or museum. Historic sites exist to remind us of our past practices and values. These same places might inspire different ways of thinking about transitioning to different energy sources. Experts all agree that human beings can mitigate climate change by changing how we use energy for heat, light, movement, and production. Interpreters need to help move the visiting public beyond the narrative that high energy use is necessary for progress. Public interpretation must expose the vast energy infrastructure and the impact of energy extraction, production, and use on places.[10] Historic sites offer place-based contexts for visitors to interact with and think critically about the processes and the impact of energy development in, for example, a maritime village. This can easily involve why and how we use certain sources of energy for certain tasks, including interpretive activities themselves.[11] Change over time is something historians—and historic sites—can and should emphasize when interpreting energy and the environment.

Unfortunately, understanding energy transition and sustainable use requires a baseline knowledge of science, which many Americans do not adequately possess. Our current political climate has further undermined confidence in scientific methods and data. Historic sites might consider mimicking what science museums now do, particularly in demonstrating potential and kinetic energy with interactive exhibits. Science museums have assumed responsibility for discussing energy's impact on climate change. The Science Museum of Minnesota has issued a statement asserting its acceptance of climate change as a scientific fact and its intent to deliver "information about global climate change as a fundamental element of scientific literacy and critical thinking."[12] At the Connecticut Science Museum, a permanent exhibit called *Energy City* focuses primarily on energy efficiency by featuring renewables, having visitors calculate their own carbon footprint, and illustrating electrical lines through a model of the capital city of Hartford. The museum website also links to curriculum guides for teachers. But energy education cannot just come from the scientific community. History museums can and should partner with museums of science, and even provide actual links to science museums.[13]

By working with and supplementing content with museums of science, history professionals can focus more on contextualizing energy in broader cultural practices that change and evolve over time. Historical sites can and should support (but offer a different perspective on energy than) science museums. This perspective should encompass the impact of energy choice and use on climate change. By providing context for past decisions and demonstrating change, historic sites can reframe and relocate the debate about energy use from media outlets, where readers and viewers may dismiss it as politics, and even from science museums, where the focus is often only on technology, to a place-based civic learning environment.

In his 1956 classic *Interpreting Our Heritage*, Freeman Tilden stressed that interpretation needs to go beyond facts and artifacts and "must be relevant to the visitor and be based in the visitors' experience." To accomplish this, Tilden advocated educational activities that involved original objects, firsthand experience, or illustrative media, rather than simply communicating factual information. Knowing one's audience and offering them specific and appropriate language, references, and examples can better engage any group, whether it is a fifth-grade class, a hiking club, or a group of engineers. Interpretation should remain consistent with an institution's mission by forging connections between the audience and the resource.

In his later years, Tilden himself expressed concern over "the environmental morass in which our boasted technology has mired us." In the 1970s, the director of the NPS tasked Tilden with developing recommendations for responding to the emerging energy crisis, which also corresponded with the bicentennial and a field-wide overhaul of interpretive practices. Tilden pressed NPS interpreters "to address the environmental crisis in their public programs."[14] He urged them to teach environmental education in a way that went beyond traditional pedagogy with a "call to action" to adult visitors as well as students, writing that "we do not tell people what they *must* do, but what they *can* do; not what they must be, but what they can be, and this by working with nature instead of against nature, and following nature's order, with man instead of against man."[15]

Today, NAI continues to support Tilden's fourth principle: "The chief aim of interpretation is not instruction, but provocation."[16] Significantly, Tilden cautioned against "preaching," and advised "provocation over instruction" to "stimulate interest" in the resources. In their works on interpreting nature and culture, Larry Beck and Ted Cable underscored the idea with their own list of principles, adding that interpretation should "provoke people to broaden their horizons" and to "encourage resource preservation."[17] In his influential *Interpretation: Making a Difference on Purpose*, Sam Ham points out that studies have proven that interpretation can encourage visitors to make personal, intellectual, or emotional connections and thus provoke them to think more deeply about their experiences, care about the resources, and ultimately influence a voluntary behavioral change. Ham's "TORE" model of interpretation provides an interpretive structure for this volume: First focus on a **T**heme to provide purpose and focus. Keep the presentation **O**rganized and **R**elevant to communicate how historical forms of energy worked. Be **E**njoyable or **E**ngaging to hold their attention. It is important to note that best practices for interpretation emphasize that "enjoyable" is different from "entertaining." As important as engaging audiences is, Ham points out that "an inherent risk of the entertainer's endgame is that the entertainment itself can steal the show, and whatever other outcome the interpreter had in mind is lost in the bells and whistles of the performance."[18] Finally, in NAI's publication *Personal Interpretation*, Lisa Brochu and Tim Merriman added a "P" for **P**urposeful to their acronym (POETRY) to emphasize a mission-driven objective that ensures that the entertainment portion of an interpretive program is purposeful and tied to the theme, in this case energy conservation and transition. NAI supports "a mission-based communication process that forges emotional and intellectual connections between the interests of the audience and the meanings inherent in the resource."[19]

Discussing historical practices and encouraging reflection around energy use can therefore address issues that cause climate change without directly debating the politically volatile issue of climate change itself. The recent American Association for State and Local History report, "Making History Matter," with its *Reframing History* communication toolkit, likewise suggests ways to encourage engagement in the process of critical thinking with a goal of "striving toward justice," in lieu of directly confronting existing beliefs that may align with political positions.[20] Whether a visitor is receptive to such self-reflection will more often than not depend upon their individual motivation for visiting the site. These reasons can range from curiosity to seeking in-depth knowledge, while others may visit for the experience, nostalgia, entertainment, or relaxation.[21] However, even if a visitor is not actively seeking out the intricacies of energy development and use, Sam Ham and others argue that one can encourage a "normative approach" that enlists people to fall in line with socially acceptable behavior through a variety of proven techniques for making even uncomfortable programs enjoyable.[22]

While Tilden and the NAI characterize the word *provocation* as inspiration, the word choice helps counter the long-held notion that historic sites must only present positive or entertaining stories. Professional interpreters stress that visitors should not have a negative physiological experience and that they should try to meet visitors' basic physical and emotional needs, but longtime museum scholars generally agree that reluctance to introduce negative or challenging interpretation to visitors has been overblown for many for years.[23] Sten Rentzhog, a promoter of open-air museums, argued that such fears delayed the interpretation of African American history at Colonial Williamsburg, for example. Rentzhog further advocated that recognizing environmental change was part of a museum's responsibility. He said that even subtle interpretations of the past in historic landscapes and elsewhere can "contribute to environmental awareness" by showing "how industrialism and consumer society have altered the environment." Just one site can illustrate "the processes leading to pollution, climate changes, and threats to the world's drinking supplies."[24]

Best practices should not lecture, enflame, or depress people, but rather empower them to leave the site or exhibit inspired and, ideally, with tools to take positive action consistent with the site or institution's mission. The International Coalition of Sites of Conscience encourages visitors to connect history to today's issues, particularly around human rights, and ideally take some kind of action. Even Tilden promoted more assertive interpretation in his later essays. In "Two Concord Men in a Boat," he argues that interpretation is meant to "wake men up" and "move them to action" and that the primary end goal of all interpretation is to persuade visitors of their obligation to protect and preserve heritage. One might argue that such heritage extends to the whole of the earth.[25]

This point about responsible and communal stewardship is key to addressing politically controversial topics like energy and the environment. Changing behavior is not necessarily tied to reasoned thought and choices. Researchers have also concluded that people tend to support energy systems with which they are familiar and encounter daily, such as those visually and historically consistent with their sense of place.[26] However, such perceptions may not always be entirely accurate, because the built environment has historically and deliberately hidden the systems that develop and deliver that energy. In our postindustrial world, energy is pervasive, but largely invisible.[27] We tend not to notice our energy infrastructure, or the impact of our energy choices. If we do see them, we do not always understand the nature of what we see. We also tend to forget why we, the consumers, have chosen to use, and choose to *continue* to use, certain energy sources over others. While in the short term we still rely on many industrial energy technologies developed long ago (in the cases of coal, oil, and uranium), we can also tap the historic uses of water, wind, and solar power for inspiration.

Interpreting energy resources can help us examine how, for what, and for whom we design energy systems. Historical interpretation can help the public see the energy systems, both historic and current, that have gradually faded from public view—infrastructure that transports energy to visitors, whether solid (coal), liquid (oil), gas, or electricity from the initial naturally occurring resource to its conversion or development all the way to consumption. Such artifacts reflect cultural values as much as technological advancement. Objects of energy are therefore artifacts worth examining.

The Material Culture of Energy

One way to educate the public about energy is to bring often-hidden energy objects and landscapes into visitors' collective visual field, allowing people to recognize, acknowledge, and identify the material culture of energy, including both renewable and nonrenewable natural resources like fossil fuels, the technological equipment of energy production, and the infrastructure that delivers and converts energy for use.[28] Whether it is the energy resource and the material scaffolding around its production and delivery, or our own use and demand as consumers, a major goal of interpretation needs to be making often invisible resources *visible*.

From 2014 until 2017, British scholars coordinated "Material Cultures of Energy," a global research project based in the United Kingdom. The grand effort, which included publications and activities, is a useful resource for identifying larger stories and themes, particularly from a global comparative perspective. One of the group's expressed goals is to target "energy communicators," including museum curators and cultural institutions, to help them use arts and humanities to educate and influence consumer choices. Number one on the list of five avenues through which to accomplish this outreach is through "object-based communication."[29] One can interpret energy through objects already located in many museums at historic sites: the products and machines that actually use the energy. These include tools, appliances, engines, manufacturing equipment, energy storage facilities, energy systems, and more. Interpretation should try to address the purpose of an item: Is it for heat? Cooking? Agricultural work? Transportation? To prevent flooding? The questions that will challenge visitors the most will be those that encourage visitors to think about how these objects reflect our cultural values, and the places people live and how.

Additionally, we can also identify and isolate organic materials like wood or coal and note their tactile, physical characteristics. Where and when was organic material first formed? Who extracted it, and how have people used it? We can look at four useful areas of object analysis: 1) Genesis, 2) Construction, 3) Function, and 4) Interpretation, and apply that to energy.[30] For example, the "genesis" is the original energy source. The universe holds a fixed amount of energy, and both human-controlled and natural processes convert that energy from one form to another. Natural resources, the technological and cultural technology of energy production, distribution, and use, are all resources that contribute to and define those processes. "Construction" includes the materials and technologies involved in processing those sources, and we can think about "function" as how we use that energy. The conversion of that object into various types of energy is part of the "construction" process. We can examine the technology that generates energy (often electricity) from resources, whether it be *potential energy* (chemical, mechanical, nuclear, or gravitational), or *kinetic energy* in the form of radiant (electromagnetic or light), thermal (heat), motion, or soundwaves. Heating, refining, and burning produce secondary energy resources that include charcoal, coke, and gasoline.

Many historic sites can illustrate energy and its production process with simple muscle energy, one of the most basic forms of energy found in humans and animals. Animals work, particularly in agriculture. Work animals like horses, oxen, and mules, as well as cows, poultry, sheep, pigs, and even dogs, provided labor, food, and cloth at rural agricultural sites. They moved materials and drove machinery. The animal is the source or genesis, and the equipment (construction) harnesses its muscle and movement into power for various agricultural and transportation tasks (function). Interpretation can examine these tasks in terms of how the animals perform the work, why they are performing it, and the significance of that work on a farm.[31] Humans built equipment to convert their activity to energy, including many types of harnesses (breastband, collar, pack saddle), yokes, and other hitching implements like tethering straps. These systems "allow efficient transfer of power from the animal to the implement or load, while allowing the animals to work comfortably and without injury."[32]

Thomas Schlereth noted that, "'Out of site' can mean 'out of sight.' Without a documented context, many artifacts remain little more than historical souvenirs."[33] Without place-based context, discussions about energy focus primarily on the science and not the human experience. The following discussions try to synthesize science with the humanities outside of popular media and other politicized spaces. They should serve as rough guides for identifying different kinds of energy resources in many historic collections or sites. Historical and cultural context explain how and why our society has made certain choices in extracting, converting, and harnessing energy, as well as the political, social, economic, and environmental impact of those choices. As humans, our dependence upon and demand for energy in the modern era threatens life on our planet, and only the public's will can diminish this threat. While cultural institutions are limited in funding and constrained by resources and audience expectations, ideally visitors should leave historical sites and museums with a sense that we have a choice when it comes to energy use. This volume supplements current calls for economic and policy changes, because as stewards of historic places, we need to do what we can in this "all hands on deck" moment to prepare for shared stewardship of our future.

Notes

1. See Sarah Sutton, *Environmental Sustainability at Historic Sites and Museums* (Lanham, MD: Rowman and Littlefield, 2015), 47, 50–51, 119–22.
2. Diane Barthel-Bouchier, *Cultural Heritage and the Challenge of Sustainability* (Walnut Creek, CA: Left Coast Press, 2013).
3. Union of Concerned Scientists, "National Landmarks at Risk: How Rising Seas, Floods, and Wildfires Are Threatening the United States' Most Cherished Historic Sites," Section 8 Report, May 21, 2014.
4. Matt Holly, "Interpreting Climate Change Virtual Course," Common Learning Portal, National Park Service, November 10, 2021, https://mylearning.nps.gov/training-courses/interpreting-climate-change-virtual-course.
5. Preservation Leadership Forum, National Trust for Historic Preservation, https://forum.savingplaces.org/learn/issues/sustainability, accessed November 1, 2020.
6. Preservation Green Lab, "The Greenest Building: Quantifying the Environmental Value of Building Reuse," 2016, https://forum.savingplaces.org/viewdocument/the-greenest-building-quantifying, accessed November 1, 2020.
7. "Preservation and Climate Change," *Preservation Connecticut News* 44:5 (September/October 2021), 4.
8. David Stradling, *Smokestacks and Progressives: Environmentalists, Engineers, and Air Quality in America* (Baltimore: Johns Hopkins University Press, 1990), 3.
9. *New York Times*, October 28, 2020.
10. Phillip Cafaro, The National Park Service Centennial Essay Series, *The George Wright Forum* 29:3 (2012), 287–98.

11. Christopher Jones observes that humanists, social scientists, and members of the public need to all be part of energy transition. Christopher Jones, *Routes of Power: Energy and Modern America* (2014), Introduction.

12. Science Museum of Minnesota, "Our Statement," https://new.smm.org/climate-change, accessed November 20, 2020.

13. "Energy City" is sponsored by *Energize Connecticut*, a private-public partner educational initiative funded by consumers on their electric bills, https://ctsciencecenter.org/exhibits/energy-city/.

14. Bruce Craig, in Freeman Tilden, *Interpreting Our Heritage*, 4th edition (Chapel Hill: University of North Carolina Press, 2008), 13–14.

15. Craig, in Tilden, 12–13.

16. Based in Colorado, the National Association for Interpretation is a nonprofit committed to advancing and professionalizing interpretation at natural and cultural sites. It sets standards and commissions, provides publications and resources, holds webinars and workshops, and issues certifications.

17. Principle 4: "The purpose of the interpretive story is to inspire and provoke people to broaden their horizons"; Principle 11: "Interpretation should instill in people the ability, and the desire, to sense the beauty on their surroundings—to provide spiritual uplift and to encourage resource preservation." Larry Beck and Ted Cable, *The Gifts of Interpretation: Fifteen Guiding Principles for Interpreting Nature and Culture* (Urbana, IL: Sagamore Publishing, 2012), 31, 135.

18. Sam H. Ham, *Interpretation: Making a Difference on Purpose* (ebook. Golden, CO: Fulcrum Publishing, 2016), 2.

19. National Association for Interpretation, "What's Interpretation?" accessed November 20, 2020, https://www.interpnet.com/NAI/interp/About/About_Interpretation/What_is_Interpretation_/nai/_About/what_is_interp.aspx.

20. *Reframing History* is a collaboration of the American Association for State and Local History, the National Council on Public History, and the Organization of American Historians.

21. In the 2007 study funded by the National Science Foundation, *Why Zoos and Aquariums Matter*, John H. Falk led a study that identified the motivations of various audiences: Explorers, Facilitators, Professional/Hobbyists, Experience Seekers, and Rechargers. See this study for details: John H. Falk, PhD, Eric M. Reinhard, Cynthia L. Vernon et al., *Why Zoos and Aquariums Matter: Assessing the Impact of a Visit to a Zoo or an Aquarium*, National Science Foundation, 2007, https://www.researchgate.net/publication/253004933_Why_Zoos_Aquariums_Matter_Assessing_the_Impact_of_a_Visit_to_a_Zoo_or_Aquarium, accessed January 11, 2021.

22. Ham, 17–19, 47, 59, 88–92.

23. Abraham Maslow defined these needs, popularly known as Maslow's hierarchy, in 1954. It includes ensuring physical comfort and a sense of safety and security, a feeling of love, appreciation, and respect for one's own knowledge, empowerment through knowledge, appreciation, and self-actualization. Douglas M. Knudsen, Ted Cable, and Larry Beck, *Interpretation of Cultural and Natural Resources* (State College, PA: Venture Publishing, Inc.), 63.

24. Sten Rentzhog, *Open Air Museums: The History and Future of a Visionary Idea* (Stockholm: Carlssons, and Ostersund: Jamtli, 2007), 237, 280–87, 325, 371, 406. The model of the "open air museum" began in Scandinavia.

25. Tilden, 182–86, and International Coalition of Sites of Conscience, https://www.sitesofconscience.org/en/home, accessed August 9, 2022.

26. Amanda D. Boyd and Amanda Miller, "Climate Change, Energy Development, and Perceptions of Place," *Human Ecology Review* 24:1 (2018), 3–22.

27. See Daniel French, *When They Hid Fire: A History of Electricity and Invisible Energy* (Pittsburgh: University of Pittsburgh Press, 2017), and Bob Johnson, *Carbon Nation: Fossil Fuels and the Making of American Culture* (Lawrence: University of Kansas Press, 2014).

28. David Wuebben, excerpts from *Power-Lined*, University of Nebraska Press, or "From Wire Evil to Power Line Poetics," *Energy Research and Social Science*, 30 (August 2017), 53–60.

29. "Material Cultures of Energy," http://www.bbk.ac.uk/mce, November 15, 2020. Funded by Britain's Arts and Research Council, four historians and a geographer examine how energy transformed daily modern life by working with a range of sources, including material objects, film and fiction, company and local archives, consumer manuals, and oral history. The researchers compare findings in Britain, North America, Germany, and Japan; Science Museum Group, "Communicating Material Cultures of Energy," https://www.sciencemuseumgroup.org.uk/project/communicating-material-cultures-of-energy-five-challenges-for-energy-communication. This paper lists five areas of communication, including: 1) object-based communication, 2) behavioral-based communication, 3) visual media communication, 4) participatory communication, and 5) community engagement. Frank Trentman, ed., "The Material Culture of Energy," *Science Museum Group Journal*, Spring 2018, http://journal.sciencemuseum.ac.uk/issues/spring-2018.

30. Andrew J. Viduka, The Protection of the Underwater Heritage, Unit 15, 2012, UNESCO (United Nations Educational, Scientific, and Cultural Organization), Bangkok Thailand, https://docplayer.net/40125217-Unit-15-author-andrew-j-viduka-material-culture-analysis.html.

31. In April 2020, the sixteenth Material Culture Symposium for Emerging Scholars at the University of Delaware focused on "Animaterialities: The Material Culture of Animals."

32. R. Anne Pearson, Timothy E. Simalenga, and Rosina C. Krecek, "Harnessing and Hitching Donkeys, Mules, and Horses for Work," Centre for Tropical Veterinary Medicine, Draught Animal Power Research and Training, May 2003, https://www.vet.ed.ac.uk/ctvm/Research/DAPR/Training%20Publications/Harness%20Hitching%20Donkeys/Harness%20Hitching%20donkeys%20Oct06.pdf.

33. Thomas J. Schlereth, "Material Culture Research and Historical Explanation," *The Public Historian* 7:4 (Fall 1985), 28; also see James Deetz, "The Artifact and Its Context," *Museum News* 62:1 (October 1983), 26.

CHAPTER 2

The Energy of the Forest

TREES ARE ONE OF OUR GREATEST WEAPONS against climate change. They absorb carbon dioxide and release the precious oxygen that humans and animals need for life. People across climatic regions have access to wood to provide energy: cooling via shade, and, of course, heating fuel. We also know from well-publicized studies that trees in the forest are connected to and interdependent on one another for the overall health of the forest ecological community, teeming with the activity of animals, birds, insects, and other organisms. For centuries, trees have also provided the primary material for shelter, furniture, tools, vehicles and vessels of transportation, paper, and food (sap, fruit, seeds). For thousands of years, most people in a community joined in the task of gathering wood for home heating and cooking. Wood stores the sun's radiant energy as chemical energy as the resulting biomatter from years of photosynthesis. Anyone who has been camping, or who has even enjoyed a backyard firepit, knows that a wood-burning fire is one of the oldest and most accessible sources of thermal energy.

When burned, wood converts that stored chemical energy into heat. While most kinds of wood can serve as adequate heating fuel, those with the least amount of sap and resin burn the most efficiently. These hardwoods include maple, oak, ash, birch, and most fruit trees.[1] Burning wood also produces secondary forms of energy for many domestic and industrial uses, such as heating water to operate steam engines or making charcoal (containing up to 90 percent carbon) from carefully controlled burning. Charcoal has fueled equipment from backyard grills to industrial furnaces.

Histories and Contexts

The editors of an energy-themed issue of Pennsylvania's state history journal in 2015 observed that "historians have yet to create literature that properly places timber harvest and iron—subjects superbly interpreted at historical sites such as Hopewell Furnace—as an important crossroads—or site of intensification—in our use of energy."[2] However, historians have addressed wood energy in recent years in publications that museum and historic site interpreters can consult. Joaquim Radkau's simply titled *Wood: A History* (2012) offers a comprehensive and global historical overview examining wood as a cultural and economic resource whose biggest reason for consumption was heat energy.[3] Probably the most ubiquitous form of energy use and thus the simplest to interpret at a precolonial, colonial, or antebellum historic site (particularly a house museum or home site) is heating. Firewood was and still remains the most common source for producing thermal energy in the home hearth or in stoves around the world, particularly for those in rural areas and the urban poor.

In his *American Canopy: Trees, Forests, and the Making of a Nation* (2012), Eric Rutkow notes that globally, access to wood dictated preindustrial settlement patterns, but American economic growth was particularly driven by wood consumption for energy. Puritan writings imply that the abundance of wood served as a selling point to encourage colonization. Rutkow provides examples of how trees, as renewable resources for both construction and energy, supported American economic development and industrial power through the eighteenth and nineteenth centuries, asserting "American attitudes toward resource consumption were formed against a backdrop of seemingly unlimited access to wood," needed to build infrastructure and for fuel.

In his classic work on New England's environmental history *Changes in the Land* (1983), historian William Cronon recounts how eighteenth-century New England colonists accessed what seemed like an unlimited supply of wood in comparison to the denuded forests of England. Many farmers reserved a woodlot on a nearby hill, making fuel accessible and simple to transport to its destination for use. Not everyone enjoyed this convenience though, and especially cold winters often led to fuel shortages and high firewood prices. A single household could harvest, chop, and burn an acre of forest a year, consuming two-thirds of it for energy. Initially, towns set aside woodlots, but as land moved into private hands, more farmers set aside woodlots on their own property. However, not only did deforestation heat and dry out the soil, but the distance and labor required for bringing fuel to hearth intensified as forests diminished.[4] "Fuel dealers" entered the wood business to meet demand and alleviate demands on individual labor, especially as colonists clustered in more concentrated settlements. Places like Philadelphia, Boston, Newport, and New York even regulated some of the pricing to protect consumers.[5]

Historian Sean Patrick Adams has examined those colonial and antebellum crises in home heating as cities and settlements in the North grew. Adams's slim volume *Home Fires: How Americans Kept Warm in the Nineteenth Century* (2014) provides a highly accessible overview of how Americans transitioned from organic wood to fossil fuel coal (specifically the low smoke–producing anthracite or "hard" coal) as a primary source of heat energy for their homes. They did so not in response to new technology, but through their gradual, and often reluctant, adoption of the cast-iron stove. The central home hearth, of either stone or brick, remained a centuries-long and steadfast cultural tradition for heat and cooking, one intertwined with familial and social life. Thus, shifting from wood to coal to produce thermal energy was not just the result of inventions by Benjamin Franklin.

Adams places the transition away from the wood-burning hearth in the historical context of economic, cultural, and urban interactions, rather than simply innovation. The hearth was hardly the most efficient method for warming the whole house, when so much heat escaped with the unwanted smoke through the chimney. That lack of efficiency required a tremendous amount of wood, further stressing places like the forested Northeast. Many English colonists considered this inefficiency as the price for health, believing in the necessity of ventilation and air circulation, instead of a closed room.[6] The limits of wood as a construction material, including flammability and durability, contributed to the rise of the iron industry. Heating, Adams points out, was one of the first areas in which the home joined the workplace as part of industrial society.[7]

Iron became one of America's most significant industries, one dependent upon wood as part of the production process. The transportation revolution allowed iron forges, which made tools, wheels, and train rails and spikes to move from rural furnace sites to sites closer to cities. Meanwhile, larger markets gained access to an array of stove technology, which burned charcoal. As an ancient, centuries-old skill, charcoal-making is associated with romantic notions of the activity and craftsmanship involved in its production. A collier turned wood into charcoal through a specific days-long technique: by heating wood in an enclosed pit or kiln—essentially baking it with limited oxygen to "carbonize" it. A charcoal pit looks like a black, smoky mound that seals the wood off from oxygen, preventing it from catching fire and burning away. The resulting charcoal pieces burn hotter and more consistently than naturally occurring wood, with far less smoke. At the same time, burning charcoal releases carbon dioxide into the atmosphere.[8]

Europeans had exploited iron ore for centuries by heating it with charcoal at sites known as "bloomeries" to produce malleable, often wrought, iron. A blacksmith could refine that product into various tools. Large-scale iron production in colonial America, characterized by the development of blast furnaces, began in the first quarter of the eighteenth century. Most of these sites produced pig iron, which a blacksmith could process further into cast or wrought iron, to make items for building and cooking such as nails, pots, and skillets. Because charcoal burned faster and hotter than either wood or mineral coal, it proved to be an essential source of energy. Iron furnaces required six thousand cords of wood (at least an acre of forestland of fuel a day) to make enough charcoal to produce one thousand tons of iron.[9]

Ironmasters looked to site their furnaces in rural areas with renewable energy sources. This meant forests, of course, but also moving water to operate the furnace. Skilled colliers converted wood into charcoal, and running water powered a wheel to operate the blowers (eventually replaced by hydraulic turbines). Blowers were the means of sustaining air flow and maintaining high and consistent temperatures in furnaces until the mid-nineteenth century. The ideal topography required a hillside to aid workers in pouring the raw materials (limestone flux and iron ore) used to make iron into the top of the furnace (known as "charging").[10] Thus, in addition to the furnace itself, a charging bridge, a cast house, a blacksmith shop, and the waterwheel rounded out an iron-making facility.[11]

England's American colonies developed a primarily agricultural economy throughout the eighteenth century. When iron furnaces were built near woodlands to make charcoal, agricultural communities grew around them, with homes, barns, stores, and workers who subsisted on a seasonal schedule. Domesticated work animals like horses, oxen, and mules as well as cows, poultry, sheep, and pigs provided supplementary muscle energy for the villages to produce food and cloth.[12] Flattened "pyramids of stone" crossed the rural countryside of the middle and northeastern colonies of Massachusetts, New York, Virginia, Maryland, Connecticut, and Pennsylvania.[13] The demand for iron during the Civil War and the simultaneous expansion of the railroad during the 1850s and 1860s temporarily sustained consistent demand for the charcoal furnace.[14] But by the end of the nineteenth century, America's economy had transformed into more of an urban industrial one than rural and agricultural. The old iron furnace sites are tied to the historic abundance of natural resources that entrepreneurs decimated. Where new energy sources like coal-fire steam allowed manufacturing and industry to relocate from rural to urban areas (and gain access to large immigrant labor pools), America evolved from a primarily agricultural to an industrial economy. When the iron and steel industries consolidated in urban manufacturing centers like Pittsburgh, Bethlehem, and Chicago, small, independent, rural enterprises could no longer compete.[15]

When British ironmakers ran low on wood for a fuel supply in the nineteenth century, they turned to coal-fire steam. Coal and coke did not crumble from heat as easily as charcoal, improving efficiency. American ironmakers, flush with wood at the time, did not innovate to that degree until anthracite- and coke-fueled furnaces in Pennsylvania adopted those fuels for refining and smelting technologies many years later. The environmental degradation that the iron furnaces caused resembled the late nineteenth-century copper pit mines of Montana or Arizona in producing industrial waste and remolding the landscape. As the demand for charcoal and heating wood exceeded the supply, colonists erected furnaces with enough distance between them to avoid competing for wood supply. Demand, not available resources, drove production.[16]

The pollution and resource exploitation of the iron-making process allows contemporary interpretive discussions about air quality and energy transition at these sites. Although not comparable to the air pollution in industrial cities, old photographs indicate emissions from the furnace and the smoke from the charcoal-making process contributed to poor air quality, especially if the furnace burned bituminous, or "soft," coal. The noises of the bustling worker villages would have been in stark contrast to the peaceful settings that the remains of the furnaces occupy today. The operations left industrial waste, known as slag, in piles nearby. Although sometimes workers could recycle slag to limit the waste, these piles are some of the earliest examples of industrial waste dumping.[17]

Only a handful of scholars have approached writing the history of the iron industry within a regional or national historical context. Historian Robert Gordon's work is probably the most comprehensive in describing the iron-making process. While his volume does not focus on the larger historical context of energy use, one looking for energy stories can certainly find abundant information.[18] The railroad competed for wood to make wrought iron as material for its expanding system. The train's engine sparks then caused forest fires, further diminishing the wood supply. Many iron enterprises then transitioned from heating charcoal to mineral coal (a nonrenewable energy) to generate steam power to run the furnaces.[19]

Gordon argues that in Connecticut, ironmakers' commitment to community, sense of place, and quality-of-life concerns kept the Salisbury ironmakers modest in their goals of increasing production and profit. In northwestern Connecticut, the regional iron industry did not try to meet the demand for wrought iron, even though Salisbury-born engineer Alexander Lyman Holley won the rights to the Bessemer steel-making process. Holley instead brought his industry to New York and Pennsylvania, while the Salisbury furnaces focused on producing high-end iron products. Gordon's explanation about the resilience of northwestern Connecticut's forests provides an example of how local culture and local control over natural resources can influence energy choices over sites of production.[20]

Interpreting Wood Energy

As the nation's earliest industrial places, the first historic sites that national and local preservationists chose to commemorate America's industrial past were iron furnaces. These sites extracted and consumed acres of forest, as much for energy as for product. Organic energy sources therefore determined the location of communities, settlement patterns, and the organization of time. Savvy businesses cropped up at these places to enlist natural resources as energy for industrial production. Iron furnaces serve as convenient places to interpret wood energy because they are located in situ to the wood that fueled the industry.

The interpretive challenge for such sites lies in our collective memory's strong adherence to Thomas Jefferson's agrarian dream of a pristine middle landscape. In contrast to this ideal, industries like mining and iron making were industrial sites set in rural areas, places that many people view today as "softer" landscapes like the "agrarian countryside or the pristine wilderness." In reality, argues geographer Richard Francaviglia, iron furnace sites became "hard places—where making a living is tough work," where business interests tried "to outwit both nature and the economy," and where workers "were constantly transforming the earth."[21] Without proper interpretation, these rural-industrial landscapes perpetuate the pristine image of rural settled areas as "the garden," an almost spiritual place located between the city and the "untamed wilderness," where independent, self-sufficient yeoman farmers carefully tended and cultivated their fields. In contrast, places with ideal resources for iron making tended to be poor, agrarian land that required sizable communities of dependent laborers.

Once defunct, however, the "working landscapes" and quaint villages that surrounded these stone pyramids elicit romanticism, not history. Amidst the literary and political pastoral ideal that had defined "the meaning of America ever since the age of discovery" is an example of a cultural paradox that scholar Leo Marx described as "the machine in the garden."[22] The physical remains of the iron industry, scattered among state parks and forests, have a devoted public following. Even while the area furnaces operated and consumed the woodlands, visitors to the countryside often did not see the conflict between rural and industry and characterized the land as a romantic rural-industrial landscape. As its forges closed, remaining residents began to repurpose, and even rename, the area to emphasize a recreational landscape that still celebrated a rural-industrial aesthetic within an "industrial ecology."[23]

These furnace sites can educate even casual, recreation-minded visitors about the long, intimate, and complicated relationship between natural resources, technology, and labor in energy production and use. While the iron furnace needed waterpower to blow air, ignite the charcoal, and maintain the heat for iron production, ironworkers used wood to make charcoal that provided the furnace's primary heating fuel. In addition, these wooded sites also needed access to iron ore for the product, of course, but also limestone flux to trigger the chemical process that separates out the molten iron. Both Connecticut and Pennsylvania produced iron for munitions and cannons, making up the primary arsenal for the American Revolution. And for roughly two centuries, workers in smoky places disposed waste into streams.

With most of the urban public disconnected from energy use in our daily tasks and products, the rural-industrial landscape of iron furnaces can convey how working people have interacted with and transformed natural resources. The modern environmental movement has emphasized man living harmoniously with nature, but in their day, the ironmaster and his workers "symbolized progress and man's domination over nature" and "how human beings have historically known nature through work."[24] William Cronon noted that only after World War II did most Americans view rural areas as "nonworking" landscapes. Historian Brian Cannon likewise observed that "as fewer Americans relied upon extractive industries for their livelihood, citizens increasingly viewed rural landscapes through lenses that privileged aesthetics, recreation, and environmental protection over extraction of resources for maximum economic growth."[25]

Figure 2.1. Author's children viewing diorama, Visitor Center, Saugus Iron Works National Historical Park, Saugus, Massachusetts, 2013. Photo by author

One of the oldest American iron facilities, the Hammersmith Ironworks (1646–1670) near Boston, first interpreted its site as a story about the grit of the Scottish prisoners of war who, as indentured servants, served as the workforce. When the National Park Service (NPS) took over the Saugus Iron Works National Historical Site in the 1970s, it shifted interpretation toward technology to emphasize themes more directly related to national significance in time for the bicentennial. An interpretative overhaul in 2001 identified themes around technology and the iron enterprise's economic value in America, but it focused secondary themes on the diversity of its workforce and the environmental impact. The only reference to energy included the "strength of waterpower" as a goal for understanding.[26]

Environmental resilience has often obscured the impact, evidence, and legacy of the charcoal-fueled furnace on these landscapes. Both Pennsylvania and Connecticut consumed so much forest for charcoal-making that, over a century later, both states emerged as leaders in American forest conservation in the early twentieth century. Today, former furnace sites are situated in areas that early twentieth-century conservationists identified for the development of "natural" preserves, public forests, and parks established to serve as respites for contemporary urban industrial workers. Interpretive programs must therefore encourage visitors to use their historical imaginations, or simulate them virtually, rather than actually re-creating these conditions. Interpretation can stress this resilience to explain that when energy needs and technologies evolve, natural regeneration can sometimes occur.

Figure 2.2.   Beckley Furnace, North Canaan, Connecticut, 2021. Photo by author

As evidenced by organizations like the Friends of Beckley Furnace in northwestern Connecticut, members of the public are passionate about preserving what remains of these resources, especially when compared to the many industrial sites located in urban places. Their efforts have opened potential for more robust interpretations through public dialogue about energy in partnership with other nearby museums and historic sites.[27] The former iron districts in Canaan and Salisbury are now pristine landscapes protected through early twentieth-century conservation efforts. At the state park–operated Beckley Furnace (built 1847), interpretive panels explain the machinery and necessarily include some information on the use of energy and energy efficiency, with particular attention to waterpower from the adjacent Blackberry River. Educational materials note the furnace used the hot-blast technique, in which workers recycle the exhaust gas from the furnace to reheat the incoming blast air, increasing production and reducing both fuel costs and the amount of carbon monoxide released into the atmosphere.[28]

At Beckley, volunteers run the interpretive program, answering questions to supplement signage that describes the resources and the iron-making process. The site's signage engages technical information in a conversational tone, and the in-person interpreters are open to further research and crowdsourcing. A waterfall adjacent to the furnace and a display of the water turbines illustrate and explain the waterpower technology. Signage frequently links the natural resources to the cultural resources, and even highlights instances of fuel and product recycling, but it stops short of explaining these operations. Expanding the

Figure 2.3.   Waterfall and turbine at Beckley Furnace, New Canaan, Connecticut, 2021. Photo by author

narrative beyond the sites' period of significance could help with stressing the Salisbury district's choice, based more on cultural and local concerns, to *not* transition to coal when steel became the preferred metallurgic product.[29]

The remains of an 1826 furnace in nearby Kent, Connecticut, stand like a sculpture against an overgrown landscape, with the historic area primarily serving as an archeological preserve. Interpretation is limited to a few text panels that explain the iron-making process and what the site might have looked like in use, including the workers and their activities. Partnerships and other site resources could allow a deeper dive into the role of natural resources in energy production. The furnace shares its site with the Eric Sloane Museum, a re-creation of the former artist's studio. The Eric Sloane Museum is a state-run site that celebrates tools, mostly wood, several of which were operated with animal muscle energy. The exhibits include both dog-powered and horse-powered threshing machines.[30] The aesthetic and environmental themes of a state historic site that celebrates labor, or at least the tools of work, struggle to convey a working landscape of continuous socioeconomic activity from energy production. Site managers have plans to clear out invasive weeds to expose the tailraces and illustrate more of the furnace complex. The privately run Connecticut Antique Machinery Museum next door features a windmill as well as a collection of large steam-combustion engines. The steam energy themes of such places, like the machinery museum, could connect to furnace sites like Cornwall Iron Furnace in Pennsylvania, where steam engines eventually replaced waterwheels to operate the blowers.

The landscape of Kent Furnace offers other interesting ways to connect its remains through an interpretive lens of environmental resilience. As in the case of Salisbury, even an exhausted natural environment can, to an extent, reclaim these places. Sloane himself advocated that the furnace ruins remain, rather than be restored, because they added aesthetic interest. Conservation practices help cultivate landscape recovery, while careful historical interpretation of the remaining cultural resources can enlist that history as a way to inspire hopeful lessons about adaptation, reuse, and resilience.

CASE STUDY: Hopewell Furnace National Historic Site, Pennsylvania

The restored historic village buildings of Hopewell Furnace National Historic Site nestle among the trees of the Schuylkill River Valley in Berks County, Pennsylvania. This bucolic atmosphere belies the sights, sounds, and smells of a blast furnace in action: dust, ash, smoke, burning charcoal, the forced air of the bellows that heated the fire, the swishing of the waterwheel, and the clanking from the smelting of iron. Iron-making operations ran intermittently for over a century, 1771–1883, manufacturing products for distant urban markets. An active, diverse, and isolated community of workers and structures grew around the glow and blast cycles of the iron furnace. While nearby farmers cultivated their soils and toiled in their fields, the ironmaster and his workers exploited the forest for fuel to make the charcoal, funneled the available water supply down sloping hills for energy, and crafted massive amounts of industrial products and implements for America's growing industrial society. As one observer noted in 1959, "The visitor today can hardly realize that the furnace—with its lazily-turning waterwheel disturbing the tranquility of this place where time has long since stopped—was once the hub of great activity."[31]

Just over twenty years prior, in 1938, the NPS designated Hopewell Furnace as the first site to earn national recognition for industrial history in the United States, but today, battlefield-weary tourists usually only visit this place as part of a school field trip or travel itinerary that includes Valley Forge and Gettysburg. Most of them arrive to take advantage of the recreational opportunities in the nearby French Creek State Park and escape from the pollution, noise, and pace of Reading to the northwest and Philadelphia, located an hour's car ride away to the southeast. Amidst the largest contiguous forest in southeastern Pennsylvania, the 848-acre park encompasses about 635 acres of woodland and 145 acres of farmland, meadows, and pastures. A haven to sportsmen, hunters, and vacationers alike, French Creek State Park, state game lands, and privately held land border the historic site.[32] Year-round visitors flock to the Hopewell area to soak in the springtime blooms, stroll under the lush shade of summertime trees, "leaf-peep" at fall foliage, and spy on the idyllic winter wonderland. At the same time, visitors can learn how industries like Hopewell Furnace extracted the very natural resources they are now escaping from the city to enjoy.

The conflict inherent in rural-industrial landscapes like Hopewell Furnace, which exploited natural resources for energy and industrial production, has plagued the park's interpretation over decades.[33] How do industrial sites stimulate critical thinking about energy in such romanticized places? Iron furnaces and other rural industries cannot go too far with industrial interpretation without losing sight of the "rural" setting and reducing much of the park's appeal to visitors. In 1993, one administrator agreed that "the beauty of the area is not true to its historic scene, when the site would have muddy areas, trees clear cut to the horizon for fuel, and be filthy with activities associated with the furnace and charcoal-making."[34]

The NPS discovered and designated the Hopewell Village National Historic Site precisely because it was located in the middle of a preserve of land that the Roosevelt administration had already set aside for an eventual state park, as a Recreational Demonstration Area based upon its rural location and former industrial use. Historic Hopewell extended past the structures and included surrounding forests and farms, but the original establishment of the historic site separated its natural resources from the cultural ones. Over the years, Hopewell's planners wrestled with the need for a museum to situate Hopewell in a larger context within the "broad picture of the iron manufacturing industry." They repeatedly considered the restoration of the natural surroundings critical to their mission and advocated the reconstruction of quarries, charcoal pits, ore and limestone stockpiles, and other iron-making facilities.[35] The state-administered French Creek State Park would put the water rights needed to operate the furnace at risk, necessitating the NPS to seek an easement and water right to Hopewell Lake to operate the furnace's waterwheel machinery. The village would lose the context of a "typical wooded area, with hearth remains of charcoal pits" to exhibit the fuel that the furnace used for operation.[36] NPS geologist Harold Hawkins advocated the installation of an exhibit describing the iron-making process in detail. "To not exploit these facts and tell the metallurgic and geologic stories," he argued, "would leave a gap in the interpretive presentation."[37]

When the 1973 oil embargo and subsequent energy shortage raised social and political awareness about energy, President Jimmy Carter called for fossil fuel reduction, and the NPS began to develop programs. In support of this type of idea, Hopewell's 1973 Interpretive Prospectus included "unnatural" exhibits like fences, outhouses, clotheslines, gardens, scattered charcoal and tools, and production waste products like wood, slag, limestone, and iron ore piles.[38] Then superintendent Elizabeth Disrude recognized the ease with which her park, one where natural and cultural resources both contributed to the site's historical significance, could comply with the energy theme. Three different interpretive programs already concentrated on the use of Hopewell's natural energy resources (wood, water, and animal power), and they could adapt their extant programs to further stress energy awareness. A contractor proposed an interpretive scheme that focused upon the furnace's energy technology, including the waterpower, iron ore, limestone, and charcoal necessary for iron production.[39]

In the 1970s, the Hopewell Furnace National Historic Site produced much of its own energy for its interpretive programs, especially in the summertime. First and most notably, the park made its own charcoal, which fueled the hearth that in turn provided the intense heat needed for the pouring demonstrations at the casthouse. It was easy enough for the costumed interpreters to raise energy awareness while demonstrating an industrial or domestic activity and compare it to what a power machine or appliance can accomplish today.

One particular and unique contemporary program around charcoal-making does more than any other to illustrate the role of natural resources in energy use, particularly the most prominent resource in the Hopewell Big Woods: the trees. With the abundance of the American forests, Hopewell consumed five to six thousand cords of wood a year for charcoal. Woodcutting, according to park materials, employed the most people. While their activity resembled clear-cutting, woodcutters who worked one part of the forest often returned twenty years later to harvest more. Hopewell's unsuccessful attempt at transition to an anthracite furnace is an important story because it marks the death of charcoal-making as a fuel source for iron and the transition to coal. Railroads preferred iron made from coke or coal, and thus mining replaced many charcoal-making jobs.[40]

Engaging visitors' senses through occasional demonstrations can be useful for showing processes like charcoal-making, a historically valued skill few have witnessed and far fewer understand how to do. For visitors, the very act of reviving the charcoal-making burns shifts their understanding of the forest from a place serving conservation and recreational purposes back to a natural resource–based place for industry, at least temporarily. It shows the conversion of wood to energy by engaging all the audience's senses. The preservation of the craft complements the molding and casting demonstrations by emphasizing the fuel necessary to generate the heat required for those tasks. When the NPS first acquired the Hopewell Furnace in 1938, historians interviewed and consulted with octogenarian Harker Long, who had managed wood production and had overseen the furnace's last blast in 1883. In the winter of 1936, when the Civilian Conservation Corps was helping to restore the site before NPS acquisition, former Hopewell caretaker and collier Lafayette Houck agreed to perform a charcoal-making demonstration to illustrate the process, which lasted several days. NPS researched and produced a detailed report and booklet on the craft.[41]

Figure 2.4. Charcoal-making at Hopewell National Historic Site, 2005. Cathy Stanton, "Cultures in Flux: New Approaches to Traditional Association at Hopewell Furnace National Historic Site: An Ethnographic Overview and Assessment," (Northeast Region Ethnography Program/ Cooperative agreement with University of Massachusetts-Amherst, National Park Service, March 2007), 199.

The Hopewell Furnace National Historic Site revived charcoal-burning in the late 1950s and 1960s, when staff recruited former Bethlehem Steel Company employee Elmer Kohl to demonstrate the process. When Kohl died in 1980, he passed on the craft to another collier to work with staff. Tasks like cutting and hauling wood for the labor-intensive, two-week burn proved too demanding for regular maintenance workers to perform with historic authenticity, and it ultimately led to an apprenticeship program in the 1990s. Each August, the park continues to hold just one annual charcoal-making demonstration, on Establishment Day, using volunteer apprentices. Public interest proved highest for the demonstrations that generated a lot of smoke, and, of course, while smoke production again creates a problem for recreational visitors, the demonstrations are quite infrequent.

In 2005, anthropologist Cathy Stanton participated in the event, covering two eight-hour shifts.[42] During what she described as originally a very isolated and rural practice, Stanton found the reenactment of the process to be fairly social, as apprentices visited and interacted with one another and tourists, even in a limited fashion. Her interviews with many of the participants conveyed their feelings of connecting to the past through this activity, and the responsibility they felt in preserving and conveying these skills to the public audiences. "People today are very much removed from their environments," commented one collier-in-training.[43]

Because the community can still suffer the environmental consequences of the historical iron making, site-specific demonstrations like charcoal-making and cast-molding programs, performed sparingly, should be continued, promoted, exploited, and contextualized for attracting visitors and demonstrating the direct process of production from energy source to product. The iron furnace landscape illustrates that natural resources *are* cultural resources through the historic iron production process.[44] The interpretive model of cultural landscapes can also help visitors reconcile the artificial separation that both local and national agencies have made in managing natural and cultural resources.

Exhibits and interpretation programs need to make more explicit connections to energy processes and their environmental impact so as to move away from nostalgia and address the site as relevant to today's concerns about ecological degradation and climate change. By 1993, under the specter of climate change awareness introduced by Vice President Al Gore, the park's new interpretive plan renewed an emphasis on energy as driven by individual choices: "While iron is important, so is change. Hopewell represents an important transition from agriculture to industry. Old transportation networks and iron-making technologies are replaced with the new. Traditional values and lifestyles are challenged by new ideas.... It is a story of a whole industry, but more importantly of individual people and how they adapt or get passed by. It is a story of the past and yet as current as the daily newspaper headlines."[45]

More recent interpretive plans have encouraged extending interpretation beyond the boundaries of the historic village and into the trail system of French Creek State Park.[46] These plans likewise confirmed the NPS's need to work in partnership with the state park to fully explore the environmental themes so recreational visitors can more fully appreciate the impact of human activity on the preserved area they enjoy today. The development of the Hopewell Big Woods, a 110-square-mile conservation area with fifteen thousand acres of broken forest dedicated to heritage and natural preservation, recognized the problematic issues surrounding boundaries at certain sites, particularly when natural resources are so integral to the operation of the cultural ones. The 2018 interpretive plan emphasized equating natural resources with cultural resources across three out of the four interpretive themes: Iron and Industry, The Land and Its Resources, and Hopewell Village as Park. It advocated a more interactive, "audience-centered" interpretation that encourages dialogue and critical thinking. This type of programming could encourage visitors to connect energy resources to industrial activity and economic growth.

Former iron furnace sites specifically provide an opportunity for visitors to directly connect industry to the natural resource extraction of wood for energy. It is therefore important to include both the natural and technological resources within the boundaries of the site. Sites with geographic limitations can form partnerships with other organizations that manage public lands like parks and forests. Beyond partnerships at the site of production, homes and urban spaces served as the market for the wood-burning stoves the iron industry produced. We can enlist material culture in the form of a furnace's products and byproducts to understand energy production, demand and consumption. Asking questions about how certain materials like charcoal or iron continue to convert energy through everyday use like cooking and heating can stimulate discussion about how consumer choice, demand, and use is part of energy history. The following are examples of illustrative artifacts that many sites and museums could acquire fairly easily, but they may need an in-person interpreter to guide visitors in their examination.

OUR $14.95 WORLD BEATER, PRICE SMASHER AND THE ENEMY OF TRUSTS

THE ACME AMERICAN RANGE.

$14.95 Buys the No. 8-18

CASH WITH THE ORDER.

WITH PORCELAIN LINED RESERVOIR AND HIGH SHELF.

FOR HARD COAL, SOFT COAL, WOOD OR ANYTHING USED FOR FUEL.

$14.95

DELIVERED ON BOARD THE CARS AT OUR FOUNDRY IN CENTRAL OHIO.

State whether you wish to burn **WOOD ONLY, COAL ONLY, OR BOTH COAL AND WOOD,** and we will send you this new big 1902 model ACME-AMERICAN 415-pound Range by freight on receipt of $14.95, and if not found perfectly satisfactory, exactly as represented, the handsomest range you ever saw, and the equal of any range you can buy elsewhere at $30.00 to $40.00, we will refund your money.

THIS RANGE WEIGHS 415 POUNDS and the freight will average for 500 miles, $1.50 to $2.00; greater or lesser distances in proportion.

THIS RANGE is made in our own foundry by skilled mechanics, from the best material money can buy, is the handsomest, most ornamental, best baking and burning and most economical big square oven, high shelf range made.

WE ISSUE A BINDING GUARANTEE

Guarantee the stove to reach you in the same perfect condition it leaves our foundry.

MONEY CAN'T MAKE BETTER. Operating our own foundry we furnish better materials, heavier castings, heavier nickel finishings, better connections and fittings than any other foundry produces. From our own factory we save you the manufacturer's, wholesaler's and retailer's profit, and give you a better range than you can buy elsewhere. Our special $14.95 price is based on the actual cost of material and labor, with but our one small profit added.

THIS BIG CAST IRON RANGE is made from the very finest Camden stove pig iron. Latest 1902 rococo molding, large square tin lined oven door, large deep porcelain lined reservoir, handsome rococo base, large high rococo shelf, heavy nickel trimmings throughout, nickel oven door panel, nickel shelf, nickel draft door, nickel tea shelf, pins, hinges, knobs, handles, etc. Duplex grate, cut tops and centers, large flues, bailed ash pan, slide hearth plate.

WE CAN ALWAYS FURNISH REPAIRS FOR ACME STOVES AND RANGES.

Prices are Cash with the Order. Delivered on the cars at our Central Ohio foundry.	Catl'g No.	Size	Size of Lids	Size of Oven	Size of Top Measuring Reservoir	Size of Fire Box when used for wood	Height to Main Top	Weight	Price	If desired without reservoir, but with end shelf, deduct $2.00. If high shelf is not wanted, deduct $2.00.
	22R375	8-18	No. 8	17½x16x11½	42x25	17x8x8	28 in.	415 lbs.	$14.95	

Oven measurements DO NOT include swell of oven door, and DO NOT include pipe or cooking utensils. See pages 580 to 593.

If you do not burn coal at all, make your order read **WOOD ONLY,** and get the exclusively **WOOD FIRE BOX** which will measure 22x9x9, but will **NOT** burn coal at all.

Figure 2.5.   There were several varieties of stoves fueled by wood, coal, or gas. Sears Roebuck and Company with introduction by Cleveland Amory, The 1902 Edition of the Sears Roebuck Catalog (New York: Bounty Books, 1969), p. 821.

Artifact Spotlight

Slag

Beckley Furnace's interpreters display examples of the raw materials needed for iron making: charcoal, limestone/quartz, and iron ore, as well as samples of slag. Slag is a necessary waste product of the process, resulting when minerals separate from the iron ore with the help of limestone flux, and is thus an essential piece of material culture. The particular makeup of those mineral impurities determined the color of the slag. Because they are in solid form, the large piles of this waste have not been particularly toxic to the Blackberry River and ecology. Interpreters refer visitors to several piles of this colorful, sometimes smooth, sometimes pocked, hardened stone, dispersed about the area. They note that ironworkers sometimes even recycled the slag for shingles or gravel. One can use slag as inspiration for interpreting how energy causes chemical reactions and material changes in natural resources that, in turn, produce physical byproducts, both useful and wasteful.[47]

Charcoal

The iron industry made and used charcoal by burning large amounts of wood in forests, but we have continued to use charcoal for other heating purposes into the present. In her 2007 ethnographic study of Hopewell, Cathy Stanton provided a succinct overview of the global charcoal trade, including the development of charcoal "briquettes" in the early twentieth century, which commodified energy with less-messy portability and easy storage. Colliers converted chipped wood into charcoal through burning, ground the charcoal to powder, and compressed the results into small squares. Those briquettes eventually powered manufacturing plants, like those belonging to Henry Ford. Today's commercially sold briquettes are currently made with wood charcoal, combined with anthracite coal, mineral charcoal, starch, sodium nitrate, limestone, sawdust, and borax.[48] Examining commercially made and sold pieces of charcoal that one can

Figure 2.6.   Slag and Charcoal Display at Beckley Furnace, New Canaan, Connecticut, 2021. Photo by author

find in one's own backyard barbecue could be a way for visitors to understand the difference between this wood-based product and the anthracite or bituminous coal that miners pull out of the ground.

Pig Iron

Furnace products allow sites not directly associated with furnace remains to interpret the industry's energy legacies by addressing the production process. Hopewell's and Beckley's operations survived as long as they did because pig iron, in demand throughout the Civil War and the construction of railroads, was the most commonly produced and versatile product of the charcoal-heated iron furnace. When the iron separated from the ore and sank to the bottom and the molten iron flowed onto the casting floor into rows of molds, it resembled pigs suckling on their mother, hence the nickname "pig iron." A blacksmith burned wood charcoal to produce thermal energy to convert these rods of "crude" iron directly into kettles, cast-iron stove parts, and even steel. Others further refined the pig iron at a forge into wrought iron to make tools, cannon, fencing, rails, and light fixtures. Many historic sites feature blacksmith demonstrations, but they do not always connect his work to that of the collier in their popular demonstrations.

Hearth and Cast-Iron Stove

Other products can go even further to help visitors understand the role of wood in creating heat. One popular iron furnace product, the cast-iron stove, transformed wood and eventually coal into thermal energy for home heating and cooking. What about the material of iron and the design of the stoves allow the conduction of heat for cooking and heating? An exhibit in Hopewell's visitors' center explained the energy efficiency of cast-iron stoves over the hearth, as well as "the rise and fall in popularity of the Hopewell plate stove and why it became desirable." The popularity of these stoves became essential to the financial sustainability of many early nineteenth-century furnaces after the demand for arsenal iron products waned. Hopewell's furnace produced four models of the Franklin stoves designed to provide more heat with less smoke.

The works of Sean Patrick Adams on early American republic and antebellum heating is the most useful in explaining how and why consumers made a difficult transition from wood to coal.[49] Howell Harris, a labor historian at Durham University in the United Kingdom, maintains a blog devoted to the cast-iron stove. Harris's research focuses on how American inventors, manufacturers, and consumers coped with the problems of exceptionally severe winters in the early nineteenth century, at a time when expectations of comfort and convenience in everyday life were growing. Abundant coal deposits and improved transport systems created opportunities to develop new technologies for heating and cooking, upon which nineteenth-century domesticity would depend. This led to the rise of large-scale manufacturing.

In the nineteenth-century kitchen, the stove eventually replaced the hearth, even in modest homes, when innovation allowed small heating stoves to also serve as baking/warming ovens, and even as stoves when one cut a hole above to heat water. Cookstoves grew larger while heating stoves shrank with the evolution of the "base burner." The wealthy, who neither needed the portability nor the efficiency of energy that the poor did, continued to separate the heating and cooking functions, eventually adopting whole-house furnaces by the midcentury.[50] By the late nineteenth century, stove production grew in areas where coal was the dominant heating source, but transportation allowed product assembly to take place in urban areas with more available labor.[51]

The coal-fire stove offered significant improvements in comfort as well as convenience, primarily because people had to spend less *caloric* energy. The stoves saved wood harvesting and cutting labor in the household. Once cooking became part of the equation, the products also replaced heating energy. Howell Harris discusses how fuel and technology shifted as the temperature in public spaces began to raise expectations for greater climate comfort in the home. Several articles, many by Harris, can be rich sources for not only discussing how people used the stoves, but what energy and economic processes allowed for the stove's development and how they, in turn, precipitated energy transition. By tracing the evolution of the hearth fire to the stove and finally to heating the entire house from a basement furnace, he offers various angles from which to discuss the stove as a material reflection of the shift from wood to coal (in anthracite and bituminous form) as a heating source from the 1820s to the 1850s.[52]

As such an example illustrates, we can interpret energy through the tools that household members, largely women, depended upon to perform tasks. The cast-iron stove saved the labor (primarily male labor) of chopping wood for the hearth. Now the female bore the sole duties of cooking and cleaning. In her study of household technology, Ruth Schwartz Cowan argues that like industrial workplaces, the same energy systems that powered manufacturing also industrialized the home. Like the manufacturing systems, she explains, "Modern housework depends upon nonhuman energy sources."[53]

Historic sites and museums with cast-iron stove collections can address energy use in tours or in printed materials. The Rensselaer County Historical Society in Troy, New York, and the Albany Institute, also in New York, feature whole stove collections. Old Sturbridge Village in Massachusetts features an online collection.[54] As material culture, stoves have attracted much attention from hobbyists and historians alike (see recommended reading). There are plentiful resources available on the stove designs themselves, but also about the innovations to those designs that enabled them to better answer needs and more efficiently burn fuel. Stoves limited smoke better than a hearth, and, most importantly, portability allowed their placement in the middle of the room for superior heating. Still, both cultural trends and economic forces of supply and demand prompted people to transition from burning wood to coal as both an industrial and a home heating source (see chapter on fossil fuels).

Some of the interpretations discussed here about energy at sites of industry can also be incorporated into historic house interpretation. Interpreters can further discuss energy use around food preparation and preparation implements, such as skillets, pots, pans, pails, wheels, and kitchen tools of all kinds. Visitors can consider other methods of heating and cooking and the cyclical nature of energy transformation and transference. For example, the stove, whose iron material depended upon the heat from charcoal made *from* wood, in turn burned wood for purposes of heat and cooking. Many stories of energy follow a cycle in this way.

Examining products and byproducts of the iron furnace can provide opportunity for visitors to engage with these artifacts as consumers, contemplate how they heat and cook in their own lives, and the types of materials they use to do so. In recent years, gas stoves and cast-iron skillets, pots, and pans have made a comeback after a trend of electric heaters, burners, and chemically enhanced, often toxic, nonstick coatings. Why? In addition to nostalgia, cast iron retains heat for a long time. The material allows one to seamlessly move dishes from the stovetop to the oven, opening more options than just frying. How the cookware performs also depends upon the source of thermal energy: gas or electric.

The stoves the iron furnaces produced played a key role in moving society into the era of fossil fuels, but a revival of wood as a renewable biofuel for home heating has received increasing interest since the 1970s for those interested in diversifying energy sources.[55] We must not forget that even iron furnaces required several types of energy to operate. They needed wood for fuel, they needed animals and humans for labor, and manufacturing also needed moving water for kinetic energy. The power of water is the subject of the next chapter.[56]

Notes

1. Eric Meier, *Wood! Identifying and Using Hundreds of Woods Worldwide*, 2008–2012, https://www.wood-database.com/book, accessed January 5, 2020.
2. Brian Black, Anne Norton Green, and Marcy Ladson, "Energy in Pennsylvania History," *The Pennsylvania Magazine of History and Biography* 89:3 (October 2015), 252.
3. Daniel French, *When They Hid the Fire: A History of Electricity and Invisible Energy in America* (Pittsburgh: University of Pittsburgh Press, 2017); Rutkow, 5–7, 347; Eric Rutkow, *American Canopy: Trees, Forests, and the Making of a Nation* (New York: Simon and Schuster, 2012), 14, 20, 23; Joaquim Radkau, *Wood: A History* (New York: Wiley and Sons, 2012).
4. William Cronon, *Changes in the Land: Indians, Colonists, and the Ecology of New England* (New York: Hill and Wang, 1983), 120–23.
5. Brooks C. Mendell and Manda H. Lang, *Wood for Bioenergy: Forest as a Resource for Biomass and Biofuels* (Durham, NC: Forest History Society, 2012).
6. This attitude toward indoor ventilation has proven relevant and prescient in the post-pandemic twenty-first century.
7. Adams has also written several articles about consumerism and energy choice in the nineteenth century that supplement this brief volume. Sean Patrick Adams, *Home Fires: How Americans Kept Warm in the 19th Century, How Things Worked Series* (Baltimore, MD: Johns Hopkins University Press, 2014), 25–45; Adams, "Warming the Poor and Growing Consumers: Fuel Philanthropy in the Early Republic's Urban North," *Journal of American History* 95:1 (June 2008), 69–94.
8. National Parks Traveler, "The Art of Making Charcoal at Hopewell Furnace National Historic Site," June 10, 2015, https://www.nationalparkstraveler.org/2015/06/art-making-charcoal-hopewell-furnace-national-historic-site26699.
9. Robert Gordon, *A Landscape Transformed: The Ironmaking District of Salisbury, Connecticut* (Oxford University Press, 2000); Gordon, *American Iron, 1607–1900* (Baltimore: Johns Hopkins University Press), 22, 34.
10. Russell A. Apple, "Mission 66 Prospectus for Hopewell Village National Historic Site" (Hopewell Furnace National Historic Site, July 27, 1955).
11. Kise, Franks, and Straw, with Menke and Menke, "Cultural Landscape Report: Hopewell Furnace National Historic Site" (National Park Service, Northeast Region, December 1997) (CLR), 24–36.

12. Gary B. Nash, et al. *The American People: Creating a Nation and a Society*, 4th edition (New York: Addison Wesley Longman, 1998), 340; Hugins, CLR, 41.

13. Much of this summary has been compiled from Joseph E. Walker, *Hopewell Village: The Dynamics of a Nineteenth Century Iron-Making Community* (Philadelphia: University of Pennsylvania Press, 1967), which provides a social framework for how rural, industrial village communities operated. Roy Appleman, "Historical Report: French Creek Area" (Bronxville, NY: Second Regional Office, National Park Service, August 19, 1935), 2. Hopewell Furnace Handbook; Hugins, 28.

14. CLR, 42.

15. CLR, 30.

16. Gordon, *A Landscape Transformed*, 14, 52, 56–58.

17. Gordon, *A Landscape Transformed*, 123, 148.

18. Robert B. Gordon, *American Iron, 1607–1900* (Baltimore: Johns Hopkins University Press, 1996).

19. David Lewis, *Iron and Steel in America* (Greenville, DE: Hagley Museum, 1976), 23–25.

20. Robert Gordon, *A Landscape Transformed*, iv, 4, 98–119.

21. Richard V. Francaviglia, *Hard Places: Reading the Landscape of America's Historic Mining Districts* (Iowa City: University of Iowa Press, 1991), x.

22. Leo Marx, *The Machine in the Garden: Technology and the Pastoral Ideal in America* (New York: Oxford University Press, 1964), 3, 73. The phrase "the garden" often referred to the paradise achieved through cultivating one's own land for use, as in the Garden of Eden; Francaviglia, 4–5.

23. Gordon, *A Landscape Transformed*, 115.

24. Francaviglia, *Hard Places*, 9. In his study of the Columbia River, historian Richard White argued that "one of the great shortcomings—intellectual and political—of modern environmentalism is its failure to grasp how human beings have historically known nature through work." Richard White, *The Organic Machine: The Remaking of the Columbia River* (New York: Hill and Wang, 1995), 4–5.

25. Brian Cannon, *Reopening the Frontier: Homesteading in the Modern West* (Lawrence: University of Kansas Press, 2009), 223.

26. Emily Murphy, "The Challenges of Interpretation at Saugus Ironworks," *FutureLearn* (Durham University, US National Park Service), https://www.futurelearn.com/info/courses/battle-of-dunbar-1650/0/steps/66292; Staff and Partners, "Long Range Interpretive Plan," Saugus Iron Works National Historic Site (US National Park Service, Department of the Interior, 2001), 6–7.

27. Two state-run sites in Connecticut, one operated by the Historic Preservation Office and the other by the state park system, serve as interesting comparisons to Hopewell. These divisions are housed in completely different agencies (the Department of Economic and Community Development and the Department of Energy and Environmental Protection, respectively).

28. Friends of Beckley Furnace, "Beckley Furnace: Industrial Monument 1847–1919," Interpretive Sign, Beckley Furnace Industrial Monument, North Canaan, CT.

29. In the video on the Beckley Furnace website, a man named Ron Jones implies that Connecticut engineer Alexander Lyman Holley wanted to preserve the landscape of northeastern Connecticut. Emphasizing Robert Gordon's thesis that residents, including Holley, made a conscious decision about preservation would be an ideal way to convey energy transition as a cultural choice. Connecticut Public Television, "Canaan Connecticut History of the Iron

Industry," 6:15, 1999, August 18, 2018; Ed Kirby, *Echoes of Iron in Connecticut's Northwest Corner* (Sharon Historical Society, 1998); Gordon, *A Landscape Transformed*, 99; Peter C. Vermilyeas, *Hidden History of Litchfield County*, supports the notion that Holley wanted to preserve the Connecticut landscape.

30. American Gnomes, "1856 horse-powered Cox and Roberts threshing machine (Kelley Farm Minnesota)," 2:45, September 2, 2016, https://www.youtube.com/watch?v=XgwOXgFaLDg; Tammy M-13, "horse powered saw at Kings Landing Historical Settlement in NB, was cool to see," 1:17, August 20, 2016, https://www.youtube.com/watch?v=RgHjfQmgoo8.

31. G. Clymer Brooke, *Birdsboro: Company with a Past Built to Last* (New York: The Newcomen Society in North America, 1959), 11. Much of the content for this case study is based upon Leah S. Glaser, "An Industrial Place in a Rural Space: The Administrative History of Hopewell Furnace National Historic Site," Philadelphia. PA: Northeast Regional Office/National Park Service/Bloomington, IN: Organization of American Historians, August 2005.

32. Kise, Franks, and Straw, 1.

33. Glaser, "An Industrial Place."

34. Derrick Cook, Superintendent, Statement for Management: Hopewell Furnace National Historic Site (October 1993), 37.

35. Robert Lee, December 11, 1939, with handwritten notes by Conrad Wirth, Matt C. Huppuch, Memorandum for the files, General Correspondence, September 12, 1938, to December 30, 1939, National Historic Sites–Hopewell, Central Classified Files 1933-49, RG 79, NACP.

36. Melvin Weig, "Report on Proposed Hopewell Village Boundary at French Creek Demonstration Project, Birdsboro, Pennsylvania" (Bronxville, NY: Region 1 District 3, National Park Service, April 10, 1937).

37. Harold Hawkins, March 27, 1941, General Files, Northeast Regional Office, Philadelphia, PA (NERO-P); Glaser, 260.

38. John C.W. Riddle, Superintendent to Assistant Director, Park Support Services, NERO-P.

39. Glaser, 229.

40. In her 2007 ethnographic report, anthropologist Cathy Stanton provided a good overview of charcoal trade around the world. Cathy Stanton, "Cultures in Flux: New Approaches to Traditional Association at Hopewell Furnace National Historic Site: An Ethnographic Overview and Assessment" (Northeast Region Ethnography Program/Cooperative agreement with University of Massachusetts-Amherst, National Park Service, March 2007), 180.

41. See John P. Cowan, "Interview with Harker Long," April 5, 1938; Motz, Monthly Report, April 5, 1941, "NMP-CCC Hopewell Village April 1, 1941, to December 1941," Box 56, RG 79, NARA-Mid Atlantic Region (Philadelphia); Jackson Kemper, *American Charcoal Making in the Era of the Cold Blast Furnace*, National Park Service, Popular Studies Series 14 (1937); Arthur Sylvester and Jackson Kemper, *The Making of Charcoal as Followed by the Colliers of the Schuylkill Valley* (Pottstown, PA: National Park Service, US Department of the Interior), January 1937.

42. Stanton, 187–99.

43. Stanton, 177–83, 214.

44. Derrick M. Cook et al., "Long-Range Interpretive Plan: Hopewell Furnace National Historic Site" (Elverson, PA: Hopewell Furnace National Historic Site, National Park Service, 1993), 6. Photocopy at HOFU.

45. James P. Corless, Roger Stone, and Jeffrey Collins, "Resource Management Plan: Hopewell Furnace National Historic Site," (1994), 2.

46. In 1961, Rogers Young reported that the park had begun burning materials in the furnace top to produce smoke, thus interpreting through smell as well as sight, but this practice might not have lasted. Staff Historian Rogers Young to Chief Historian, June 27, 1961, K1817, "Interpretive Prospectus," Central Files, HOFU; Denver Service Center, "Master Plan," 1972, 7; Glaser, 292–93; Hopewell Big Woods Trails and Recreation Concept Plan (National Park Service, Northeast Region, Rivers, Trails, & Conservation Assistance Program and Natural Lands Trust for the Hopewell Big Woods Partnership, October 2009), 42; The Hopewell Big Woods Partnership, "Hopewell Big Woods," http://hopewellbigwoods.org/about.html, accessed July 1, 2021.
47. Friends of Beckley Furnace, Inc., "Beckley Furnace: Learning Resources: Archival Material," https://beckleyfurnace.org/resources/archival-material, accessed July 1, 2021; Gordon, 168.
48. Stanton, 180–82.
49. Adams, *Home Fires*.
50. Ruth Schwartz Cowan, *More Work for Mother: The Ironies of Household Technology from the Open Hearth to the Microwave* (New York: Basic Books, 1985), 52, 58, 52–64.
51. Howell Harris, "The Architecture of the American Stove Industry," 4, https://docs.google.com/document/d/1qf_B85gr4YsRVaSoLNED6W1diQHKrDubbpae--0ZpYo/edit, accessed September 26, 2021; Howell Harris, "Stove History Stuff," https://sites.google.com/site/stovehistorystuff/home/early-research- personal web page, accessed September 26, 2021.
52. Howell Harris, "Coping with Competition: Cooperation and Collusion in the US Stove Industry, c.1870–1930," *Business History Review* 86(4): 657–92, 2012; Howell Harris, "Inventing the U.S. Stove Industry, c. 1815–1875: Making and Selling the First Universal Consumer Durable," *Business History Review* 82(4): 701–33, 2008.
53. Cowan, 6.
54. "Cast Iron Stove Production at Hopewell Furnace," Hopewell Furnace National Historic Site, Pennsylvania, National Park Service, https://www.nps.gov/hofu/learn/historyculture/cast-iron-stove-production.htm, accessed July 28, 2021; Robinson and Associates, "Hopewell Furnace National Historic Site" Historic Resources Study" (December 1, 2004), 35; Howell Harris, "A Collection of Stoves from American Museums, I: Plate Stoves," Blog, October 22, 2013, http://stovehistory.blogspot.com/2013/10/a-collection-of-stoves-metropolitan.html, accessed July 28, 2021; Howell Harris, "Articles and Chapters," https://sites.google.com/site/stovehistorystuff/home/articles, accessed September 26, 2021, Old Sturbridge Village, Online Collections, http://resources.osv.org/explore_learn/collection_list.php?G=16&S-G=44&A=BI, accessed July 28, 2021.
55. Brooks C. Mendell and Amanda H. Lang, *Wood for Bioenergy: Forests as a Resource for Biomass and Biofuels* (Durham, NC: Forest History Society Series, 2012).
56. National Park Service, "Hopewell Furnace: Hopewell Furnace's Waterwheel," https://www.nps.gov/hofu/learn/historyculture/hopewell-water-wheel.htm, accessed February 26, 2015.

CHAPTER 3

Water Is Life, Water Is Power

IN 2017, THE PROTEST AT THE Standing Rock Indian Reservation against the Dakota Access Pipeline rallied around the indisputable tenet that "Water Is Life." In addition to disturbing Indigenous sacred sites, the pipeline threatened to disrupt and potentially poison the water table, upon which the Lakota community depended for drinking. Native people have protested energy development for decades, including power generated by water at such sites as Rainbow Bridge in Utah. Such places have intense spiritual and religious meaning. All human evolution and survival have required a water source. Control over, and even ownership of, water has dictated social and economic power throughout much of human history. Moving water has carved out rock in places as vast as the Grand Canyon. Its geological impact shaped and continues to dictate the landscapes and define the places in which we live. As a source of nourishment, transportation, and energy, water serves humans in multiple ways.

Throughout the eighteenth and early nineteenth centuries, New England's waterways attracted both agricultural settlements and industry. Water transported goods from numerous rural industrial sites to population centers. It provided power for grist and fulling mills, sawmills, tanneries, and iron furnaces. Mill operators likewise built dams along rivers to divert and store water in wet seasons. When captured, control of moving water greatly increases its economic value. Some dams simply harness the energy from the streamflow, but many dams double as storage and allow controlled water release. Release from that supply allows them to control and take advantage of waterflow in times of drought. Iron furnaces needed wood for fuel, but also moving water for kinetic energy. Headraces and tailraces moved the water in and out of the waterwheel.[1]

The waterwheel became and still remains an intimate part of the idealized rural-industrial (also known as "working") landscape for not just iron furnaces, but also mills and forges. The wheel moves in accordance with the speed of the streamflow and the height from which the water is falling. For hydroelectricity, water released through a large pipe, or penstock, leads to a turbine that powers the generator where a rotor with electromagnets converts the mechanical energy to electricity. Because neither communities nor individuals often use energy immediately, the more efficiently we produce it, the more we need to store it for another time when we need it. *Storage* is therefore an essential piece of energy production, and therefore a cultural byproduct. For water energy, this may mean a reservoir and dam. These hydroelectric facilities serve a dual purpose. They store the water for domestic, municipal, and agricultural use, but when they release it, the force of the falling water rotates the turbines that produce electricity.

Landscapes with rivers, creeks, and waterfalls became valuable and coveted sites for producing hydropower: kinetic and later electric forms of energy on demand. These places, however, are not discrete. They are all connected by a network of tributaries and watersheds. This interconnectedness makes them challenging to interpret without a multidisciplinary approach that recognizes how cultural values have altered landscapes and geography to create energy and disrupt ecological systems.

Histories and Contexts

Several classic scholarly works examine the history of human control over water, for irrigation as well as energy. Donald Worster's *Rivers of Empire* (1985), Marc Reisner's *Cadillac Desert* (1986), and Donald Pisani's *To Reclaim a Divided West* (1992) are important works on this theme in the western region. John T. Cumbler's *Reasonable Use* (2001) examines the economic, social, political, and environmental impact of water control on the industrialization of an increasingly urbanized Connecticut River Valley. Other histories focus on case studies, with Patrick Malone's *Waterpower in Lowell* (2009) providing an accessible review of waterpower for manufacturing.[2]

Water is also an important component to energy systems: the iron furnace water wheel, heating for steam, heating and cooling for atomic reactions, leaching in the mining of coal and uranium, or hydraulic fracturing for oil and gas. Water is essential for growing energy crops like corn for ethanol, a biofuel.[3] However, stories around hydroelectricity encompass themes of not just technology, labor, and urbanization, but also environmental justice and ecological harm. In *The Organic Machine* (1995), Richard White explores the natural relationship between humans and the environment by examining how people have harnessed and used the energy of the Columbia River in the Pacific Northwest. This instance is only one example of many around the world that can show how perceptions about nature and technological progress can reap economic and social benefits, but also have dire effects on the entire ecosystem.[4] Hydroelectricity has transformed entire regional economies and lifestyles, but it remains difficult to transport, even coupled with long-distance transmission. We can deliver coal and steam to congested urban areas where factories and workers are, but only the development of long-distance transmission technology (see later chapter) made hydroelectricity an efficient and clean source of industrial energy.[5]

One can find primary sources about water in the archival records, whether at local historical societies, libraries, or even municipalities and utilities. Historian Donna J. Rilling identified water sites across Pennsylvania through county surveys, supplemented by contemporary newspapers, from the eighteenth and nineteenth centuries.[6] Ann Durkin Keating's handbook *Invisible Networks: Exploring the History of Local Utilities and Public Works* (1994) is particularly helpful for framing and researching local history through water systems as well as gas and electric.[7] Even if your site does not focus or utilize waterpower directly, water access and water systems are integral to most stories of community and economic development.

Interpreting Waterpower

Several types of historic sites offer venues for emphasizing mechanical or electrical energy generated from moving water. The efficiency of "white coal" (a nickname for hydroelectricity in the early twentieth century) with its themes of conservation, technology, urbanization, and industry are greatly tempered by the profound consequences to ecologies and violations of environmental justice. These consequences touch all segments of society—social, political, economic, and cultural—including community displacement and the destruction of sacred places. Natural waterways (particularly rivers), dependent upon how users up- and downstream used and facilitated or interrupted continuous streamflow, defined regional energy systems and the development of adjacent communities.

The National Park Service has conserved several rivers for interpretation under the designations of "National Heritage Corridors" and "National Wild and Scenic Rivers." Rivers have historically hosted thousands of transportation vessels that use various types of energy for propulsion, including muscle, wind, steam, and diesel. In the southeastern United States, the Muscle Shoals National Heritage Area (affiliated with the National Park Service) interprets the landscape around the Tennessee River and its impact on the region's cultural development, including of the Tennessee Valley Authority (TVA), a New Deal program that promised to "modernize" the region with hydroelectricity. Interpretive responsibilities are shared with the University of Alabama's public history program. As an interdisciplinary field, public history is equipped to interpret the cultural themes within a social-studies educational frameworks to discuss issues like energy. An interpretative program can situate a story along a broader energy timeline, within social and political context, or explicitly link hydropower to economic growth. An example of the last would be connecting the produced hydroelectricity to regional industries.[8]

The Connecticut River Museum in Essex, Connecticut, features the river itself as the primary "artifact" upon which interpretation is based. The Connecticut River provided the energy for multiple industries along the valley from Springfield, Massachusetts, to the Coltsville armory in Hartford, shifting east and emptying into the Long Island Sound at New London. The location of the museum on the riverbank adjacent to "Steamboat Dock," and occupying a rehabilitated boathouse is key to generating immediate interest in the museum content. In the late 1990s, one of the state's primary electric utilities sponsored a utility-funded exhibit, a theme they might consider reviving in some form due to the pressing need to better educate the public about energy. The exhibit, called *River Power*, traced the history of how residents harnessed the Connecticut River for power, first with waterwheels to drive grist and sawmills and eventually to produce hydroelectricity, and later as part of a larger nuclear power system for cooling reactors. While it deemphasized the environmental impacts of harnessing that energy for power the otherwise fairly text-heavy exhibit benefited from the inclusion of more hands-on opportunities, similar to those featured at large urban science museums, including a working waterwheel and models of a dam and canal system.[9] Several museums, large and small, have enlisted similar exhibits to illustrate waterpower. The following three show a range of different approaches and goals that share site-specific and interactive strategies.

CASE STUDY: The Eli Whitney Museum and Workshop, Hamden, Connecticut

The Eli Whitney Museum and Workshop is located on the Mill River, which flows into the Long Island Sound at New Haven. The historic site is situated near the southernmost end of a consequential and highly significant precision manufacturing corridor extending southward from Springfield, Massachusetts. The Mill River powered more than fifteen industries in the nineteenth century. The people indigenous to the area, the Quinnipiac (People of the Long Water), enlisted the river to transport supplies, collect fish and oysters, and maintain agricultural villages. Mill stones offer evidence of the site as a corn-flour mill as early as 1640 until the operators sold the facility to Whitney in 1798. The enterprise was adjacent to a short ridge of trap rock between the Mill Rock and East Rock, whose ridges partially dammed the river. The resulting waterfall and others like it powered the earliest of American industries, which developed a system of manufacturing that arguably set the United States on its path of global economic prominence. The site's evolution is a story of energy transition in manufacturing.

Most students of history know Eli Whitney as the inventor of the cotton gin, which he designed to remove the seeds and husks from the raw cotton fiber more efficiently than human labor. The machine's increased efficiency ironically contributed to the expansion of plantation slavery, maximizing human muscle energy along racial lines across the South. After realizing little profit from the cotton gin due to patent disputes, Whitney established a firearms-production factory at the site of the former gristmill. There in New Haven, he streamlined and amplified the energy of human labor (primarily gunsmiths) with a system of manufacturing involving interchangeable parts. On May 13, 1798, he wrote to Yale friend and then Secretary of the Treasury Oliver Wolcott: "I am persuaded that machinery moved by water ought to do the business which greatly diminishes labor and facilitates the manufacture of these articles. . . . There is a good fall of water in the vicinity of this town which I can procure and have works executed in short time."[10]

The prominent place of this quotation on the museum walls, website, and marketing materials implies waterpower as a primary interpretive theme. The Mill River's kinetic energy provided the power to operate machinery, plus the streamflow to transport charcoal downriver to be stored in charcoal storage sheds (the remains of which are some of the

Figure 3.1.   Waterworks at Eli Whitney Museum and Workshop with dam spillway in background, Hamden, Connecticut, 2021. Photo by author

oldest resources on the site). The unpredictability of the river's flow, including occasional freezing, prompted Whitney to build a dam and spillway to supplement the waterwheel and ensure a reliable energy source. The dam and millpond (the reservoir that supplies potential energy for a mill) supplied power to the forge, through two channels, with a tailrace returning the water to the river flow.

When son Eli Whitney Jr. assumed management over the armory in the 1840s, competitors were installing steam power turbines, so he installed an impulse hydraulic turbine wheel housed in iron, increasing the factory's efficiency by 40 percent. After rebuilding the factory in the 1860s, the six-foot dam his father had built still did not produce the consistent power his new enterprises needed. Whitney Jr. constructed a larger stone dam allowing for a stronger flow. He took advantage of New Haven's need for a reliable water supply, with the city helping to fund the new dam and reservoir. The move also limited other industrial development and competition elsewhere along the Mill River. The second highly productive armory made Samuel Colt's firearms, which became popular and iconic across the American West and even Europe. By 1888, the Winchester Repeating Arms Company in New Haven took over the patents and productions at the Whitneyville site. Eventually, steam and electric motors replaced the waterpower, but the dam, waterfall, and reservoir remain today.

Today, the former factory sits at the foot of the dam and roaring spillway that still supplies water to the city of New Haven, not beside a lazily turning waterwheel as it did in the mid-nineteenth century. The historic site's main function is K–12 education, but it did not begin that way. Historical markers embedded in the masonry of the dam itself discuss Whitney's cotton gin and rifle manufacturing contributions rather than the site or the river itself. On the eve of the national bicentennial in 1974 and its designation as a historic landmark, the New Haven Colony Historical Society commissioned a study to turn the armory remains into an outdoor museum and exhibit.[11] In his 1995 classic book of essays, historian Mike Wallace noted that deindustrialization had inspired a rise in popularity for museums of technology and invention.[12]

A bicentennial-era study of the Eli Whitney Armory examined historical and economic interest in the site at the local, state, and national levels. The consultants advised that it had the greatest potential as a nationally significant historic site to illustrate the full scope of America's early manufacturing process, including energy production. Conveniently located in a beautiful setting at the base of East Rock, the site had visible and accessible physical resources. Beyond restoration of the buildings, recommendations for interpretation included archeological excavations, a multimedia presentation of the site's history with photographs and primary sources, an artifact exhibit of the Whitney milling machines, examples from the Winchester gun collection, and other manufactured items. Notably, to address energy, the site would feature live demonstrations by instructors, specifically "the existing water turbine, constructed by Eli Whitney, Jr. will provide the power from the Living Museum, a reconstruction of the metalworking shop of the late 18th and 19th century. Waterpower will work the bellows of the forge and drive the primitive trip hammer and lathe."[13]

Practical matters undermined those recommendations. The site offers limited visitor access, and its location in a flood plain compromises its suitability to hold collections. More importantly, its value as a site for local urban water supply needs prioritized that use. Still owned by the New Haven Water Company, the site continues to not only provide the water to the local area, but with an electrical substation on the property, it also continues to deliver energy. The museum nevertheless opened with its first exhibit to the public in 1984, but by 1995 all traditional exhibits stopped, with site managers concluding that "heritage tourists are not the market to support the site."[14]

Eventually, the site's mission developed into a unique type of experiential history museum, one where mostly local visitors and students participated in the experiential demonstrations inspired by Whitney's style of innovation. After acquiring a collection from A. C. Gilbert, a New Haven toy manufacturer, the museum coalesced around interpreting ingenuity and process, not historical context or narrative, to honor Whitney's legacy. Today, the museum barely mentions firearms manufacturing, often to the confusion of casual out-of-town visitors, in favor of general invention and innovation themes (the Whitney "legacy") that stress hands-on learning as a supplement to the classroom and emphasize STEM (Science, Technology, Engineering, Math) skills over those of the humanities.

William Brown, the museum's director for thirty years (retired in the summer of 2021), intimately tied the museum's interpretation to replicating Whitney's process and "workshop" to honor the legacy of the site, if not the specific historical context. He focused on school-aged children as the primary audience and hands-on learning as the principal pedagogy. A statement explained that "in 1993 we adopted a dual mission of preserving the Whitney legacy—the material legacy of the site and its products . . . and the legacy in spirit—the tradition of Yankee Ingenuity. Yankee Ingenuity is the tradition of invention we pursue."[15] They interpret stewardship as the preservation and cultivation of process and innovation over artifacts.[16] Creativity can certainly be cultivated, but through process, rather than example.[17] Michael Webber convincingly argues for this kind of STEAM approach (adding "A" for Art), noting that only with innovation (and some childlike wonder and clarity) can we solve our energy challenges as we strive for clean and efficient energy production and consumption.[18]

The museum's primary programming includes a popular summer day camp and interactive afterschool programs in technology and innovation. Interdisciplinary themes broadly include the process of invention, stories of people and their times, projects that measure and encourage mastery through competition, and seeking global perspectives on solving problems. Students engage their senses and problem-solving skills by experimenting, manipulating, and creating products using water, wind, light, magnetic and mechanical forces, gravity, sound, motion, and electricity. They construct waterwheels, design various water vessels to test in the water laboratory at the base of the waterfall, and build solar-powered toys. Science and children's museums often feature similar activities, but this site has the potential to meet the goals of historical interpretation, as well.[19]

This unique model incorporates a highly competitive paid apprenticeship program that engages teens with professional designers, engineers, artists, and artisans to conceive and develop design projects for younger students. At least one project focuses on the waterwheel and, as mentioned, several enlist work with solar energy. However, both teachers and apprentices admit they could emphasize energy use more and draw explicit parallels to the historical resources and context onsite.[20] Primary sources might also supplement project programming, but those would need to be fairly brief to stay within the time and scope of the workshops and camps.

With programming running year-round, site visitation numbers in the thousands every year. Because of the popularity of the educational programs, workshops, summer camps, and the public park known as East Rock Park, the site attracts a more diverse and multi-disciplinary audience than other small historic museums, offering tremendous educational opportunity for examining the many forms of energy through what physical resources remain. The strong role of waterpower and the use of charcoal at the site invites interpretation about the role of energy in locating places of human labor and manufacturing, including exploring various forms of energy resources and their uses.[21] The museum has continued to steward the properties. Interested parties can find a good deal of important historical information on the website, but the schoolchildren, as well as the apprentices, are not engaging with this history in their activities at the site.

However, just in the last few years, the museum director produced interpretive signs to distinguish the historic site from the recreational use of the nearby East Rock Park. While not part of the programming, the recent outdoor signage has more explicitly offered a useful, highly accessible public site of historical energy interpretation with which visitors and parents can engage.[22] One notable panel about the newer, more efficient water turbine points out that waterpower is a renewable energy and focuses on the turbine that Whitney Jr. developed. The interpretation notes that electricity did not replace waterpower at the site until the 1930s (but doesn't explain why).[23]

Another panel titled "Transition" emphasizes historical change as different industries took over the site. Interpretation and programming at the site can be developed around energy transition through signage, but through artifacts, as well. Acme Wire produced wires there for telegraphs, telephones, radios, motors, and electrical distribution and transmission lines until demand outgrew the site. The Sentinel Stove company built natural gas-fueled stoves that, by World War I, encouraged people to transition from coal and woodfire stoves for cooking food and heating homes and workplaces. In 1918, an inventor took over the site and operated it similar to how Edison's Menlo Park was operated.[24] This allowed the site's continuous use as a place of innovation, the theme that most dominates the museum today, featuring inventors beyond Eli Whitney himself to discuss figures like Renaissance genius Leonardo DaVinci and A. C. Gilbert (famous for toys that provided hands-on learning). Programming opportunities around renewable energy, which they already have, can perhaps incorporate more locally focused historical interpretation. This would guide children in thinking about the drawbacks and benefits of new technologies and engage them in thinking about the circumstances that historically led people to make choices and changes in how to access and use energy.

Institutional partnerships could enhance this programming and the site's interpretive opportunities. Except for an aging interactive, outdoor wooden water exhibit the museum designed in partnership with the Water Authority and the National Science Foundation in 1993, the historical connection between the Eli Whitney Museum and the South Central Connecticut Regional Water Authority has been diluted in local collective memory. The Regional Water Authority continues to extend further guardianship to the river, which is so hidden by overgrowth and restricted by wire fencing that few local residents have a relationship with it or, in some cases, even an awareness. Plans to renovate the historic dam are underway as of 2022. The museum and water authority could partner with local public history programs and consultants to develop curriculum to train teachers who bring their classes to the site for hands-on learning. In 2021, the museum began collaborating with the statewide advocacy group Preservation Connecticut on interpretation and historic preservation programming that would teach schoolchildren older construction techniques, as well as the environmental benefits of recycling buildings.

CASE STUDY: Lowell Mills National Historical Park, Lowell, Massachusetts

While the bicentennial-era interpretive study in the 1970s identified the Whitney armory as a major potential tourist destination, the model of the industrial "Urban National Cultural Park" emerged farther north. Lowell Mills National Historical Park is the premier National Park Service site for making waterpower a major interpretive theme. The park's Suffolk Mills exhibit portrays the hydroelectric power production necessary to operate the mills most vividly.

At least one ranger-guided walk, the Suffolk Mills Walking Tour, reviews how the canals and turbines worked. The tour itself draws connections to the Merrimack River, previously used for transport, which powered the mill technology through a sophisticated canal system, but the information shared, as in most tours, depends on the knowledge of the guide. Notably, this walk includes the environmental impact of the steam and hydroelectric plants all along the river, which highlights energy choices and transition. Due to its distance from the main part of the site at the Boott Museum and Boarding House, Suffolk Mills is unfortunately not regularly staffed. Because it must be part of a tour, casual visitors will often miss it. The *River Transformed* exhibit is therefore located farther away from the other resources, which are downtown in the original power center of the Suffolk Mills complex. It features a restored operating Francis hydraulic turbine, which served as a model turbine design worldwide for decades. Water pours from a thirteen-foot drop from a penstock into a rotating turbine, which turns the gears, belts, and pulleys.[25]

An interdisciplinary approach to interpretation that explores engineering and waterpower through both science and history makes the canal and waterpower system integral to the historic significance of the site, in addition to the labor story, which attracts the most public interest. Labor, the mill's economic growth, and the industrial city depended upon the river flow and its kinetic energy. Cathy Stanton explained in her book *The Lowell Experiment* (2006) that planners deliberately wanted Lowell to focus on workers over technology and the capitalists. National Park Service staff redesigned its turbine exhibit in 2006 to emphasize the manipulation of the natural environment in service of business goals. Lowell's nineteenth-century industrialists exploited energy from both muscle and water. Interpretation about the muscle energy expended by workers to produce textiles should emphasize the connection to the kinetic energy provided by waterpower.[26]

Much like the Eli Whitney Museum, the educational STEM programs at Lowell titled "Power to Production," "Engineer It!" and the virtual program "Waterpower: Powering a Revolution" (in association with the University of Massachusetts-Lowell) invite children (grades three through twelve, depending on the program) into the museum to examine waterpower and make those connections. They workshop through hands-on experimenting,

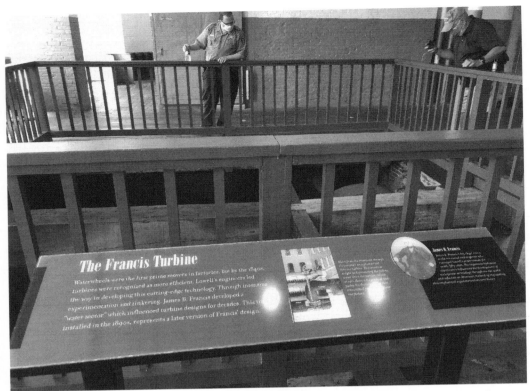

Figure 3.2.   Suffolk Mills turbine exhibit, Lowell National Historical Park, Lowell, Massachusetts, 2021. Photo by author

but then make explicit connections with the historic resources with "behind-the-scenes" tours of the Suffolk Mills' working turbine, the Moody Street Feeder Gatehouse, and then experiencing how that power production would operate machinery in the vast loom room. After the tour of the site's resources, the "Power to Production" program brings students into a laboratory to test waterwheels, design a mill and canal system, and manipulate pulleys and gears to understand how that waterpower translated into mechanical energy to operate the looms. Three different pre- and postvisit lessons (for grades four through twelve) on potential and kinetic energy can reinforce waterpower concepts.[27]

Interpretation targeting the broader public takes place outside the buildings. The Lowell National Historical Park extends throughout the city's downtown, featuring several more outdoor tours that are routed past artifacts of historic technology (operational and defunct, although preserved in place). The "Waterpower Walk" features turbine pits, lockhouses, and dams. The self-guided tours include interpretive signage and public art. Unlike traditional exhibits, they reach both park visitors and those who may be more or equally as interested in outdoor recreation.

Even more broadly, sites like Lowell can extend interpretation to the power system as it extends throughout the larger region up and downstream as Theodore Steinberg describes in his book, *Nature Incorporated: Industrialization and the Waters of New England* (1994).[28] Extending interpretation beyond park and municipal borders will take more collaboration, but it can be an important step for educating a larger audience about the interconnectedness of regional power systems and their continued reliance on natural resources.

In Phoenix, Arizona, the regional utility company brought historical interpretation to the public, integrating both art and history into its water and power system that extends throughout the Salt River Valley. In the 1990s, Dolores Hayden had appealed to public historians to find the "power of place" through public art on the streets of cities, even when the landscape has few historical resources left. Years later, Nancy Dallett issued "A Call for Proactive Public Historians," in which she advocated taking historical interpretation and education into the streets and among the resources "to reach people informally as they walk, drive, relax in parks, and engage in the fabric of their communities."[29] The following case study in Arizona's Salt River Valley, on which Dallett worked as a consultant, therefore offers a model.

CASE STUDY: The Salt River Project by Leah G. Harrison, Manager, Research Archives and Heritage, Salt River Project, Phoenix, AZ

The Salt River Valley's earliest canal system was built by ancient desert dwellers beginning around AD 500 and operated for more than a thousand years. Beginning in the late 1800s, farmers and businesspeople expanded and upgraded the valley's network of canals and laterals into the system that the Salt River Project (SRP) operates today. SRP is a community-based not-for-profit water and power utility serving the Salt River Valley and

Figure 3.3. Arizona Falls, Phoenix, Arizona. Salt River Project, Phoenix, Arizona, https://media.srp-net.com/photovideo-galleries/, accessed February 15, 2022

greater Phoenix Metropolitan Area since 1903. SRP provides electricity to more than a million customers and delivers around 800,000-acre feet of water annually, while operating a federal reclamation project comprised of a water storage and delivery system that includes eight dams and 1,300 miles of canals and laterals. As the valley has grown and developed, the changes have dramatically altered the visual landscape. SRP piped and modernized once ubiquitous waterways as the greater Phoenix Metropolitan Area has urbanized, and the organization continues to serve the vital purpose of delivering water across the valley.

Due to its vital role in the region's history, SRP has recognized the importance of historical consciousness and has engaged in public history practices in both its internal and publicly facing operations. The organization has interpreted the history of water and power in the valley for decades through an on-site museum space, publications, presentations, its corporate website, and outreach programs. In more recent years, this interpretation has migrated out of these traditional spaces and into the community at the physical sites of water conveyance and power production. Today, SRP's sites of interpretation include signage, public art, and mobile digital platforms. SRP combines art and history to educate the public on the past, present, and future of the system SRP operates.

Partnerships are critical for bringing these projects to fruition. The waterways that SRP manages are owned by the federal government and flow through numerous valley cities. The partnerships between SRP, the U.S. Bureau of Reclamation (USBR), and local municipalities allow the public to experience and understand the rich history of the valley's waterways. One result of these partnerships is historic interpretive signage across the SRP canal system. These signs, designed to evoke the visual of gates that open and close to control the flow of water, are scattered across the canal and lateral system. The content of the signs interprets the history of the place in which they are situated. This technique immerses the reader in the story of a particular landscape and connects people to the use of infrastructure over time—sometimes extending back to the Valley's earliest Indigenous inhabitants—making visible a landscape that is often no longer present.

Cities in the Salt River Valley view the canal system SRP operates as an additional amenity, turning this infrastructure into opportunities for public art, education, recreation, and transportation corridors. Working with SRP, cities have paved over eighty miles of canal banks, enhancing regional connectivity. Often these pathway projects include art and signage as additional educational recreation amenities. For example, the Scottsdale Waterfront, a unique location along the Arizona Canal, hosts Scottsdale Public Art's event Canal Convergence. This internationally acclaimed ten-day event features innovative art installations in and along the banks of the canal.

Arizona Falls, a small hydroelectric facility on the Arizona Canal, is a prime example of this intersection of infrastructure, art, and history. A granite outcropping created a waterfall at the site during the construction of the Arizona Canal in the 1880s, which became a community gathering place. In 1902, the site transformed into a hydroelectric generating facility before ceasing operations in the 1950s. In the early 2000s, the City of Phoenix, SRP, and the USBR partnered to revitalize and reinterpret the site. The new hydropower facility, designed by renowned Boston artists Lajos Héder and Mags Harries, showcases art, poetry, and technology. It was designed as a community recreation spot, allowing the public to enjoy the cooling mists of the falls as they did a century before.

The upper level of the facility includes an area referred to as the Stoa Deck. The concrete floor of the deck is sandblasted with lines from Arizona Poet Laureate Alberto Rios and imprinted with reeds. The northeastern end of the deck is inlaid with boulders sourced near the dams SRP operates along the Salt River. The lower level of the site is comprised of the water room, featuring waterfalls on three sides and boulders to allow visitors to sit and experience the roar of falling water. Pieces of the site's original generating station are mounted on the walls of the water room.

In the mid-2010s, interpretive signage was added on-site, along with a web-based virtual tour. This interpretation expounded on the history of Arizona Falls, while providing insight into the operation and maintenance of the valley's canal system—explaining the source of the water flowing through the canal system, how canals have been cleaned over time, and the various jobs and individuals that play a vital role in water delivery. Arizona Falls represents the most extensive interpretive project on SRP's system to date and is open to the public and utilized for tours and events.

More recently, SRP's interpretive efforts have embraced more digital platforms. Utilizing the ESRI Story Maps platform, SRP launched a Heritage Map in 2018. This virtual map provides interpretation and photographs of historic water and power infrastructure across the Salt River Valley. Additional layers feature information about sites relevant to the history of the valley, along with photographs, information, and locations for public art projects across the canal system. The map also details the location of interpretive signage along the canals. This is the first time we have mapped public art and interpretive signage across the system. The Heritage Map is designed to work on mobile devices and includes a location feature to allow users to see what sites of interest are nearby to learn more about the history of their neighborhood or community.

Interpretation of the valley's rich history and the legacy of water and power development has taken many forms in the nearly 120 years of SRP's operation. While traditional modes of interpretation remain important, this messaging has moved out to the actual sites of delivery or production themselves, connecting the public to the history of a particular place. As technology evolves, utilizing virtual or digital platforms will further enrich this onsite interpretation and provide additional opportunities to tell the story of SRP and the valley.

Cities outside the United States have also integrated the historic and current

Figure 3.4. Interpretive signage from Canal Project (Marie Jones, design; Nancy Dallett, Text), Salt River Project, Arizona. Salt River Project Archives, Phoenix, Arizona

waterpower infrastructure into the modern city landscape. The Ottawa Electric Power Company in Ottawa, Canada, historically interprets its municipal hydropower system at Chaudière Falls as a historically significant place dating back to its role as a sacred site for First Nations people, in addition to a way to promote "green energy." Public tours include the powerhouse of Canada's oldest operating hydro station. While it has been updated, original dials, meters, and switches are preserved in place to acknowledge the historical significance.[30] Museums can bring some of this exterior interpretation inside with the aforementioned equipment as artifacts, but also with artwork. The power of art is especially compelling with waterpower sites that tapped into and encouraged cultural values about the relationship between water, energy, and industry.

Artifact Spotlight

Landscape Paintings

Arguably the most important interpretive sign at the Eli Whitney site features a painting by William Giles Munson of the rural industrial village of Whitneyville that "is an image that finds its way into textbooks that describe the beginning of the Industrial Revolution in part because it recalls a beginning of friendly human scale and dignity not yet darkened by smoke." The text explains that the Munson painting (1826) "shows industry that is still rural and in comfortable harmony with nature," but one might invite visitors to further unpack that statement to discuss just how industry altered the landscape and harnessed energy for production.[31] One can use these types of images as primary sources to help visitors think more critically about how images have conditioned people both in the past and today to perceive waterpower.

Munson's famous painting, like the landscapes of Thomas Moran and Albert Bierstadt from the Hudson River Valley School, shaped public memory and perceptions of the relationship between water, energy, and industrial places as natural. As with the iron furnace, the images are misleading. They imply that energy and industrial production used natural resources without a negative impact, either aesthetically or ecologically. We can enlist the art and photography that propagates Leo Marx's *The Machine in the Garden* paradox to inspire conversations about how images influence our attitude toward and use of natural resources. A guide might ask visitors to articulate the ideas of harmonious nature and industry that such an image conveys alongside similar ones, such as depictions of puffing steam trains cutting through western mountain ranges or steamships along the rivers of the Midwest mass produced by companies like Currier and Ives.[32]

Paintings like Paul Starrett Sample's *Norris Dam* (1935) reveal a far different, but just as celebrated, western landscape. The government often hired artists and photographers to document large public works projects. The artwork often shows technology absorbing and controlling the natural landscape. In the twentieth century, the federal Bureau of Reclamation would construct numerous storage and hydroelectric dams through the very landscapes that Moran and Bierstadt painted almost a century before. Such projects celebrated the triumph of modern technology and communicated the efficient use of natural resources for the greater public good of sustained western settlement.

Figure 3.5. William Giles Munson. *Eli Whitney Gun Factory* (1827). Yale University Art Gallery, New Haven, CT

In the 1960s, the Bureau of Reclamation commissioned artwork of water storage sites, spillways, power plants, and even switchyards. Many celebrated the technological designs as art, but Norman Rockwell's *Glen Canyon Dam* offered overt visual commentary. Interested more in people than objects, Rockwell painted the dam and multiple power lines, and blatantly raises serious questions about the impact of such development on Native American communities. A museum located in a region that receives energy from these structures can use these paintings to press visitors about what these depictions imply about energy use, technology, and environmental justice (see more in electricity chapter).[33]

Waterpower Equipment

By 1840, coal and steam engines replaced water sites on the working landscape as well as in water and land transportation. Those historic waterpower sites, in the form of waterwheels, foundations, or still operating turbines and power plants, represent technology eligible for preservation as historic resources. While some of this infrastructure debris may be abandoned, a lot of waterpower equipment is still functional and remains an integral part of the same communities the water sources first attracted and then fostered. These still-operational systems are integrated with both older and newer technologies. Equipment often requires technological updates or unit replacement. With the passage of the National Historic Preservation Act in 1966, several agencies such as the Bureau of Reclamation and the Federal Energy Regulatory Commission have entered programmatic agreements to address issues of resource preservation. While engineers must often replace equipment or parts, the tedious historical documentation required has produced voluminous resources for interpretation, whether through National Register nominations, Historic American Engineering Record reports, or in accordance with the agency's Historic Properties Management Plan. Such efforts are meant to preserve energy and water history, and not to "impede the safe and efficient production of energy."[34]

Waterpower sites around the world remain integral parts of the industrial landscape. They are ideal venues for people to understand the use of natural resources for energy by encountering history on a daily basis. This has become increasingly difficult after September 11, 2001, when cities feared terrorist threats on their water sources and either obscured or cut off the resources from the public. We need to make this infrastructure visible at historic sites, but also recognize that water is part of the landscape and perhaps that is where interpretation will be most effective.

Notes

1. National Park Service, "Hopewell Furnace: Hopewell Furnace's Waterwheel," https://www.nps.gov/hofu/learn/historyculture/hopewell-water-wheel.htm, February 26, 2015.

2. Donald Worster, *Rivers of Empire: Water, Aridity, and the Growth of the American West*, reprint edition (New York: Oxford University Press, 1992); Donald Pisani, *To Reclaim a Divided West: Water, Law and Public Policy* (Albuquerque: University of New Mexico, 1992); Marc Reisner, *Cadillac Desert: The American West and Its Disappearing Water* (New York: Penguin, 1993); John T. Cumbler, *Reasonable Use: The People, the Environment, and the State, New England, 1790–1930* (New York: Oxford University Press, 2001); Patrick Malone, *Waterpower in Lowell: Engineering and Industry in Nineteenth-Century America* (Baltimore, MD: Johns Hopkins University Press, 2009).

3. Michael Webber, *Power Trip: The Story of Energy* (New York: Basic Books, 2019), chapter 1 (ebook).

4. Richard White, *The Organic Machine: The Remaking of the Columbia River* (New York: Hill and Wang, 1995).

5. Thomas Hughes, *Networks of Power: Electrification in Western History 1880–1930* (Baltimore: Johns Hopkins University Press, 1993; reprint). Also see Toni Rae Linenberger and Leah Glaser, *Dams, Dynamos, and Development: The Bureau of Reclamation's Power Program and Electrification of the West* (Washington, DC: US Government Printing Office, 2002).

6. Donna Rilling, "Locating Philadelphia's Water-Powered Past," *Pennsylvania Magazine of History and Biography* 139: 3 (October 2015), 356–59.

7. Ann Durkin Keating, *Invisible Networks: Exploring the History of Local Utilities and Public Works* (Malabar, FL: Krieger Publishing Company, 1994).

8. Video Conversation with Carolyn Crawford, April 7, 2021.

9. Steve Grant, "A Sense of History Flows from River Museum Exhibit," *Hartford Courant*, October 1, 1998; "River Power," Press Release, May 22–November 15, 1998, Connecticut River Museum, Essex, Connecticut.

10. Davis, Cochran, Miller, Noyes Architects, Guilford, "Preservation of the Eli Whitney Gun Factory Site and its Potential Development as a Historical Site Museum" (prepared for the New Haven Colony Historical Society); November White, *The Organic Machine: The Remaking of the Columbia River* (New York: Hill and Wang, 1995, 1974), 1, Hamden Historical Society, Miller Library, Hamden, CT.

11. Davis et al.

12. Mike Wallace, *Mickey Mouse History and Other Essays on American Memory* (Philadelphia: Temple University Press, 1995), 75–100.

13. Carolyn Cooper and Merrill Lindsay, *Eli Whitney and the Whitney Armory* (Eli Whitney Museum, 1980), Hamden Historical Society, Miller Memorial Library, Hamden, CT.

14. William Brown, phone interview with Leah S. Glaser, July 6, 2021.

15. Report, William Brown and Sally Hill, "The Eli Whitney Museum: A Case Study," c. 2002. Copy in possession of the author.

16. Brown and Hill, "A Case Study."

17. Eli Whitney Museum, "About Us: Why We Do It," https://www.eliwhitney.org/7/about-us/why-we-do-it, accessed August 30, 2021.

18. Webber, 249–50.

19. Brown and Hill, "A Case Study;" William Brown, phone interview with Leah S. Glaser, July 6, 2021.

20. Interview with Jonah Heiser, Whitney Workshop apprentice, August 18, 2021; phone and in-person conversations with William Brown, Ryan Paxson (July 27, 2021) and Andrew Sargent, Eli Whitney Museum (July/August 26, 2021).

21. Based on conversations with Andrew Sargent, Ryan Paxson, and William "Bill" Brown, August 26, 2021.

22. Press Release, "Eli Whitney Museum Opens to the Public on September 15!" September 9, 1984; "The Birth of a Museum: Eli Whitney and the Eli Whitney Armory," New Haven Colony Museum, September 5–October 15, 1978; Clause Solnik, "Museum opens window on the works of Eli Whitney Site," *Hamden Chronicle*, October 3, 1984; Davis et al.; Carolyn Cooper and Merrill Lindsay, *Eli Whitney and the Whitney Armory*, Eli Whitney Museum, 1980, Hamden Historical Society, Miller Memorial Library, Hamden, CT.

23. "The Turbine," Interpretive sign, Eli Whitney Historic Site, Hamden, CT.

24. "Transition," Interpretive sign, Eli Whitney Historic Site, Hamden, CT.

25. Lowell National Historic Park, National Park Service, "Suffolk Mills Turbine Exhibit," https://www.nps.gov/lowe/learn/historyculture/suffolk-mills-turbine-exhibit.htm, accessed September 1, 2021.

26. Cathy Stanton, *The Lowell Experiment: Public History in a Post-Industrial City* (Amherst: University of Massachusetts Press, 2006).

27. University of Massachusetts-Lowell, "Tsongas Industrial History Center, Field Trips," https://www.uml.edu/Tsongas/Education-Programs/Field-Trips, accessed September 15, 2021.

28. See Theodore Steinberg, *Nature Incorporated: Industrialization and the Waters of New England* (Amherst: University of Massachusetts Press, 1994).

29. Nancy Dallett, "A Call for Proactive Public Historians," in Rebecca Bush and K. Tawny Paul, eds. *Art and Public History* (Lanham, MD: Rowman and Littlefield, 2017), 159.

30. "Experience Chaudière Falls," accessed January 30, 2022, https://chaudierefalls.com; Chaudière Island—Ottawa Electric Power Houses Walking Tour, April 17, 2013, National Council on Public History, Ottawa, Canada.

31. Eli Whitney Museum and Workshop, https://www.eliwhitney.org, accessed August 5, 2021.

32. Also see Donald C. Jackson's book on the visual culture of dams. Donald C. Jackson, *Pastoral and Monumental: Dams, Postcards, and the American Landscape* (Pittsburgh: University of Pittsburgh Press, 2013).

33. "American Artist and Water Reclamation," United State Bureau of Reclamation, https://www.usbr.gov/museumproperty/art/index.html, accessed September 1, 2021.

34. FERC and Advisory Council on Historic Preservation, "Guidelines for the Development of Historic Properties Management for FERC Hydroelectric Projects," May 20, 2002, 11.

CHAPTER 4

The Cultural Power of Steam Energy

THE PREVIOUS CHAPTER DISCUSSED the quaint imagery of the waterwheel in the American imagination, but few sources of energy rival steam in sentimentality and popularity. Some of the most popular experiences for visitors in cultural tourism like steamboats and steam trains are associated with the coal-fired steam engine. Steam engines were economically critical, not only to the global industrial revolution in factories, but also for transporting goods and people. The prosperity, economic impact, and sense of adventure associated with steam transportation ensured that many people (and whole communities) attach powerful and positive memories, associations, and emotions to the technology.

Histories and Contexts

The steam engine allowed for portable and on-demand energy by heating water to its gaseous state. Steam technology is a secondary energy resource, since it is produced from a primary resource such as wood, oil, or most popularly, coal. The stationary steam engine converts heat to mechanical and electrical energy. Steam pushes and pulls pistons to pump water or, thanks to innovations by James Watt, to operate machinery and vehicles. As previous discussions indicate, humans burned fossil fuels like coal for millennia, but primarily for heat—thermal energy, not kinetic energy. Using coal to produce steam power helped create enormous advances in industry and transportation, and the steam engine helped transform the world economy from one dependent on wood and water to one primarily based on nonrenewable fossil fuels.

Steam power additionally served as the primary means for transporting people, goods, and materials. While the steam engine proved to be too heavy and bulky for small modes of transportation like the automobile, trains and water vessels had room to house the coal and accommodate its weight. Many historians credit Connecticut's John Fitch with the first operational steamboat in 1786. About fifty years later, coal replaced wood as the primary energy resource to fuel steam and steam-powered boats. Locomotives ignited the transportation and market revolution in the 1820s, and by the 1840s, most steamboats had turned to coal due to diminishing wood supplies.

The historiography of steam energy has primarily focused on its technological history. In his prizewinning and classic *Steamboats on the Western Rivers* (1949), Louis C. Hunter detailed every aspect of riverboating from engineering structure to passenger life, but he stressed their economic role as carriers of commerce across the expanding United States. Even earlier, Henry Dickinson published *A Short History of the Steam Engine* (1939), but Carroll Pursell's *Early Stationary Steam Engines in America* (1969) compared the American and English technology during the Industrial Revolution. However, Pursell's main argument is that the steam engine was vital in antebellum America for manufacturing in places where the water supply was unpredictable or unreliable—specifically in the Old Northwest/Midwest, rather than in New England. R. Douglas Hurt's *American Farm Tools: From Hand-Power to Steam-Power* (1982) offers an especially accessible and useful resource for museums with agricultural tool collections. He reviews the technology, but also examines why and how knowledge, need, and affordability all determined the shift from muscle to steam.[1]

In 1989, Richard Hills provided a technological history of the stationary steam engine, *Power from Steam,* that could serve as a guide to the engine as material culture. However, many historians turned their attention from technological history to the economic impact of steam power. Maury Klein's *The Power Makers: Steam, Electricity, and the Men Who Invented Modern America* (2008) and William Rosen's book *The Most Powerful Idea in the World: A Story of Steam, Industry, and Invention* (2010) contextualized steam within the framework of innovation and invention. Rosen's focus on how technology propelled the adoption of burning fossil fuels is certainly critical to understanding climate change.[2]

As widespread concern for climate change grew, so did scholarly attention to cultural and social associations around the history of steam. In 2016, Andreas Malm placed steam history squarely within the framework of climate change, drawing interdisciplinary, direct, and unapologetic causal lines in *Fossil Capital: The Rise of Steam Power and the Roots of Global Warming*. Malm focused his questions on the nontechnological circumstances that compelled capitalists' energy transition from waterpower to steam power. Steam power became more popular because it could increase production. And just as significantly, steam power transformed transportation via land and sea and expanded national and global markets.[3]

Interpreting Steam Power

Historic interpretation of steam power has tended to emphasize the relationship between capitalism, economic power, and the dominance of fossil fuel energy. In addition to transportation and maritime museums, heritage trains and steam engine museums around the world preserve working steam-energy technologies. Many of these resources do not operate in their original setting. This would normally be problematic since historic preservationists have long stressed the importance of setting and context for understanding a resource. Isolating artifacts of this type, however, allows easier access for larger audiences and educational opportunities through sights and sounds.

Many museums offer occasional demonstrations, in part to aid in maintenance and ensure the preservation of the steam resources, which benefit from occasional use and operation. "Engineeriums," like the New England Wireless and Steam Museum in East Greenwich, Rhode Island, celebrate innovation. This museum holds the largest collection in the country of stationary steam engines, a technology that brought the small state national recognition. The museum hosts an annual "Steam Up" where it practices an over-half-century tradition of operating the machines all at the same time.[4] While such demonstrations are exciting, engaging, and can attract hundreds of visitors, they can obscure the original fuel for these engines. They also miss an opportunity to think critically about energy.

For those sites where artifacts with coal fire steam engines have remained in their original or similar setting (such as trains and steamboats), preserving the technology has clashed with concerns for air quality and threats of climate change. Environmental interests have challenged the operations and preservation of popular, site-based historic vehicles at two well-known historic sites: Mystic Seaport and the Durango and Silverton Narrow Gauge Railroad. These challenges brought complex interpretive questions into the public arena. The resolution of these preservation questions offers intriguing models for interpreting energy in a way that balances the public's desire for historically authentic and nostalgic experiences with rapid technological and climatic changes. Provocative interpretation can invite public reflection and conversations around both the historical and present significance of energy.

CASE STUDY: The *Sabino*, Mystic Seaport Museum, Mystic, Connecticut

One of the last wooden, coal-fired, black smoke–emitting passenger steamboats on the East Coast has served as a popular attraction since 1973 at the Mystic Seaport Museum, one of the state's premier tourist attractions. It shuttles passengers along a river excursion while regularly releasing black plumes of smoke. While staff acknowledged that many visitors might view the boat ride as merely entertainment, museum materials stressed that the emphasis of the trip would be educational for the roughly one hundred passengers the *Sabino* could carry at a time. Other research materials insisted that only a working, interactive exhibit could accurately convey "the smell of hot lubrication oil, the clocklike sound and motion of the engine linkages, and the grating of the coal on the scoop as the engineer spreads the coal on the fire."[5]

The privately funded effort to preserve Connecticut's maritime history with a living historic seaport village gave Connecticut its own maritime version of Williamsburg and Hopewell Furnace in 1929. Like Sturbridge Village in nearby Massachusetts, Mystic Seaport gathered historic buildings and watercraft at the site to mimic a nineteenth-century maritime community. Both mission and nostalgia motivated museum board members to bring a steamboat to Mystic to join its restored vessel collection as an interactive exhibit to carry museum visitors as passengers on river cruises. In the early 1970s, trustee Henry

Figure 4.1. The *Sabino* with coal, Mystic Seaport Museum, Mystic, Connecticut, 2011. Photo by author

DuPont observed that steam power was not represented at the museum. Furthermore, "An operating steamboat could provide visitors with an incomparable understanding of the power and mechanical intricacy of the age of steam. It would allow them to experience history rather than just hear about it."[6]

The growing popularity of living history and experiential learning museums in the years leading up to America's 1976 bicentennial celebration provided additional motivation to pursue an interactive exhibit. Mystic Seaport's Master Plan stated that outdoor exhibits could "provide the visitor an opportunity to participate in and experience a sense of 19th century life by exposing him to a three-dimensional environment which interprets aspects of New England's maritime community." "Encouraging active participation is one of our goals and *Sabino* is a means of achieving it," staff asserted. "Being on board while the boiler is being fired or the engine is responding to bell signals from the wheelhouse, and of course, when the steam whistle blows, is a unique event for most young people and we hope it will be an educational one as well. Older visitors for whom steamboats were once a common means of coastal transportation will recall pleasant memories."[7] Due to its smaller size, the *Sabino* proved to be "the most logical approach to our interpretation of steam," since other ships were simply too big to navigate the Mystic River.[8]

Excursions were consistent with *Sabino*'s historic use. In the mid-nineteenth century, railroads replaced the waterway as a trade route, encouraging more leisure-oriented river trips.[9] The *Sabino* outlasted the other small coal-fired "steamers" of its kind. Several different entities had owned and operated the boat since it was built in 1908, restoring and altering its original form as needed. In 1941, the Casco Bay Line to Portland Harbor converted all its steamboats to diesel *except* for the *Sabino*, which passengers preferred to the noise of most diesel engines. When the Corbin family rebuilt the ship (following an accidental sinking), they added a viewing area for passengers to observe the engine operations, replaced the boiler, lengthened the smokestack for more efficient operation, and added staircases for additional safety. The family believed this would appeal to the most passengers, asking, "Who wants to watch a noisy, stinky, diesel engine?"[10] From 1967 to the early 1970s, the *Sabino* chartered tourists along the Merrimack River in New Hampshire until the Mystic Seaport Museum assumed the steamboat on loan for a year to test it as an interactive exhibit on the Mystic River.[11]

To the museum, *Sabino*'s historical significance hinged on its status as the sole-surviving, still-operational, small-excursion steamer on the Atlantic coast with its original engine.[12] By the time the steamer arrived on the Connecticut shoreline, the owners and operators of *Sabino* had altered the form of the steamboat over time because transportation technology, like all technology, is constantly evolving to meet new needs and advances, which thus raises preservation questions. Managers decided to represent *Sabino* as she was when she arrived, but its condition necessitated large-scale restorations to become operational.[13] When the National Register of Historic Places listed *Sabino* in September 1991, the form's preparer downplayed the alterations, explaining that the ship remained in "excellent condition and would be recognizable to those who knew her in service in the early years of this century." The nomination form highlighted certain features, one being the original Paine steam propulsion engine (rebuilt in 1991), for qualities that characterized the vessel as the type consistent with a basic structure despite changes in its upper deck and canopy.[14] *Sabino*'s steam engine remains fully exposed for visitors to observe in full operation. Running the engine required up to seven wheelbarrows (120 pounds each) of coal. A day's steaming would burn eight hundred to a thousand pounds of coal fuel. Workers needed to perform the messy task of shoveling out ashes from the firebox. The museum stored the coal in a shipyard bin for loading into a port and starboard bunker.[15]

The timing of the decision to acquire the *Sabino* proved somewhat problematic. The 1969 National Environmental Policy Act, followed by the first Earth Day in 1970, boosted a growing environmental movement that included the passage of a national Clean Air Act (1970) and Clean Water Act (1972). The political fallout from the 1973 Yom Kippur War disrupted worldwide oil supplies, further fueling the environmental movement's emphasis on alternative energy. Within this historical context, the *Sabino*'s own history as a maritime exhibit provides another critical interpretive opportunity to think about steam as a sustainable source of energy and why it is no longer the most efficient or prudent one to use today, even when it serves as a historical exhibit itself.[16]

After it had been operating a few years, Williams College student Sarah Willis reported in a class paper, now preserved at the Mystic Seaport Library, the challenges of stewarding an interactive exhibit to interpret steam technology. Questions from Connecticut's Department of Environmental Quality (DEQ) arose within months due to *Sabino*'s use of bituminous coal, which produced plumes of heavy, dark smoke. The museum felt it could not consider a new fuel source as an option, because, precisely due to the many physical alterations to

Sabino over the years, the original 75-horsepower Paine compound two-cylinder engine and its coal fuel source remained the most historically significant feature of the vessel. According to their 1973 annual report, Mystic Seaport hired an environmental lawyer recommended by Oliver Jensen, founder of the nearby Essex railroad, who faced similar conflicts with environmental regulations stemming from the Clean Air Act. Through Jensen, the seaport secured cleaner burning coal from the Pocahontas Mine in West Virginia.[17]

Additionally, Connecticut's first DEQ commissioner Daniel Lufkin excused the boat's sulfur dioxide emissions that exceeded regulatory levels (*Sabino* burned coal at .63 percent rather than regular .5 percent). Lufkin deemed the boat suitable "as an operating museum piece" after the museum secured a variance from the regulation by proving to the DEQ that 1) the extra amount of sulfur would not endanger public health and safety, 2) compliance with the regulation "would produce practical difficulty or hardship to the museum without equal or greater benefits to the public," and 3) "it would not interfere with any relevant surrounding air quality standard."[18] The DEQ continued to issue the variance annually, provided the museum would 1) still look for other suitable coal, 2) continue to investigate the possibility of converting to other fuel, 3) notify the commissioner if there is an increase in coal consumptions, and 4) only blow the whistle, which uses steam, as required by the Coast Guard. The state agency exempted the museum from meeting air quality standards precisely because changing to diesel would "nullify her historical authenticity and value" to the museum.[19]

Depending upon the visitor's preference, the *Sabino* can be an active, interactive, or leisurely experience. Passengers aboard the *Sabino* get to observe the machinery in operation. In addition to the open engine room that allows riders to see the steam operations, a myriad of sensory stimulation help visitors associate peaceful and nostalgic memories with the *Sabino*. One older gentleman "missed the creaky sighs of an up and down steam engine, the scrape of the fireman's shovel on steam plates, the loud clank of the furnace door, and the proud stentorian-bass blast of a steamboat whistle. Also, the red-hot iron, and black sooty coal smoke, and the temple of the straining steamboat as she pulls from other dock."[20]

The experience of the excursion competed with the educational value of preserving the technology. While the meditative aspects appealed to visitors, the educational content the *Sabino* passengers took away remained unclear. Passengers reportedly avoided the engine room in the summer, but they jockeyed for seats next to the warm engine in the cooler months. The *Sabino*'s captains recorded tourist comments in the early years, which implies that the excited passengers were either not listening, unaware of the *Sabino*'s historic value and integrity, or simply distracted by the excursion itself. According to logbooks, one visitor remarked (after watching the coal fire in the boiler), "Good show, what really runs her?" Others inquired, "How do you burn coal in a diesel?" and "Where do you put the key to start the engine?"[21]

The *Sabino* does not technically violate state and federal air quality standards, but the visual perception and the romanticized experience, while entertaining, could do more to educate visitors about energy production and transitions by including some of the boat's own history regarding air quality. Part of the reason for justifying the need for the steam engine was the valuable opportunity to learn about the transition from sail to steam technology.[22] Passengers have access to information on the boat from various sources. The museum has printed a brief history on the tickets, which passengers get to keep as souvenirs. An interpretive plaque near *Sabino*'s dock described the boat's history and technological significance, without including details about the large coal bin on the boarding dock. The boat ride itself runs more like a cruise, with the interpreter explaining and pointing out sights along the river and shoreline and talking about the ship, its history, how she works, and the development of steam power. Interpretive materials surrounding the ship focus on steam technology and the machinery of the engine, boiler, piston, vales, gauges, and cranks, rather than much about the coal. Staff and passengers alike consider shoveling the coal, the smell of it burning, and the smoke, as all part of the experience.[23] Nostalgic experiences can entice people, but also stifle opportunity to use the past to inform the present.

In the last two decades of the twentieth century, the media began highlighting scientific studies warning about greenhouse gases and global warming. As visitors and neighbors grew more sensitive to environmental issues in the early 1990s, Mystic Seaport looked to address some of the negative perceptions about *Sabino*. In a magazine article, the writer commented that "clouds of black smoke and soot billowing from the stack may be a part of steamboating which is gone forever. While nostalgic to some, coal, dust soot, and smoke is offensive to others in our environmentally sensitive world of today." To reduce smoke levels up to 75 percent, the operators found a historic solution in a 1909 engineering manual about smoke reduction. They shifted the jet steam horizontally into the lower part of the stack, increasing turbulence and draft in the firebox, and inducing more complex combustion of particulates and gases.[24]

While the museum staff hear some complaints from neighbors on the river, fewer report them to the Department of Energy and Environmental Protection (DEEP), which comes out occasionally to check the *Sabino*'s emissions. One source at Mystic Seaport reported that a handful of people complain a year to the museum, and DEEP similarly claims that it has only received seven complaints for the Essex steam train and five complaints for the *Sabino* since the early 1990s. The steamboat captain tries to avoid shoveling the firebox when opposite certain private boat owners on the river, and the museum has made efforts to minimize the soot by scrubbing the decks and keeping the paint washed and the brass polished.[25]

Only a few educational materials at Mystic challenge students to consider how people historically perceived the transition from steam engines to diesel, which was cleaner but noisier. A local schoolteacher developed a curriculum around *Sabino* that asks students to consider perceptions of the technology by having them look at how passengers described their steamboat experience versus a similar diesel-powered ship. Most hated the smoke, but they appreciated the quiet. The lesson asks questions about stewardship and interpretation, including: Should museums enlist artifacts like the *Sabino* so that visitors can experience the past, or should they be preserved unused for the future? What do these objects tell us about the changes in our society? How do museums and museum objects like *Sabino* help us to evaluate the benefits and challenges of science and technological change in the twentieth century?[26] Reviewing Mystic Seaport's decisions about resource management and *Sabino*'s interpretation as part of this lesson could really engage visitors on important questions about how energy resource stewards can balance responsibility to historical accuracy with the imperative of the climate crisis (especially when its historical status exempts it from regulation).

Figure 4.2.   The *Sabino*, Mystic Seaport Museum, Mystic, Connecticut, 2011. Photo by author

In recent years, Mystic Seaport has embraced climate change and sustainability as part of its mission to preserve and interpret maritime history. In doing so, the museum has confronted questions about messaging and green technology. While building an award-winning LEED-certified exhibition building, interpreters acknowledged the *Sabino*'s smoke and asked visitors to imagine an entire river full of such vehicles. In 2018, Communications Director Dan McFadden tried to explain that only an exhibit like the *Sabino* can provide people with an authentic experience and understanding of coal-fire technology, but also about the environmental impact "and thus encourage the promotion and use of alternative power solutions today."[27]

The public's growing sensitivity toward environmental issues, and the challenges of operating historic technology, have prompted visitors to think more about our use of energy. After another lengthy restoration from 2014–2017, *Sabino*'s stewards realized that they had to either replace the boiler or just adopt an electric engine to replace the coal-fire system. The reality that captaining a steamboat is a rapidly disappearing trade added to the burdensome decision; the museum could not train enough captains to meet the demand for excursions, further encouraging the transition. Apprenticeship programs, requiring three years for certification and licensing, are specific to historic vessels like this one. They are time intensive and hard to sustain, so recruiting and retention continued to be a challenge limited to this particular boat.

With electric engines now small and lightweight enough to replace the space and weight of the coal, vice president of watercraft preservation and programs Chris Gasiorek announced in 2019 that the museum would add an electric propulsion engine with lithium batteries, motors, and control panels to supplement the coal-fire system—its weight and size can replace what would normally be pile of coal. For the purpose and authenticity of the *Sabino* exhibit, the steam power engine will stay preserved, maintained in place, and used on special occasions. Unfortunately, after several rounds of approval, the Coast Guard, which must regularly approve the *Sabino*'s "seaworthiness" and regulates the number of passengers allowed and where they stand on the deck, ultimately denied the addition of lithium batteries. The reason is likely related to studies in late 2020 and early 2021 that cited fire dangers. Staff quickly shifted to plan B: an electric motor powered by two diesel-fueled generators. These will suffice until the historic vessel can carry a safer technology. The solution is not ideal, but it allows the *Sabino* to perform multiple excursions in the interim without actively burning coal. The museum worked to adapt the hull of the *Sabino* to support the new electric diesel engine.[28]

Mystic's decision to transition the energy source of its historic steamboat to an electric motor, while maintaining the ability to use the older technology, could open a valuable theme for interpretation at the shipyard. An electric motor maintains the quiet, auditory experience of the steam engine, while the elimination of smoke could become an opening

Figure 4.3.   Work to add the electric diesel motor to the *Sabino* began in the summer of 2022. Mystic Seaport Museum, Mystic, Connecticut, 2022. Photo by author

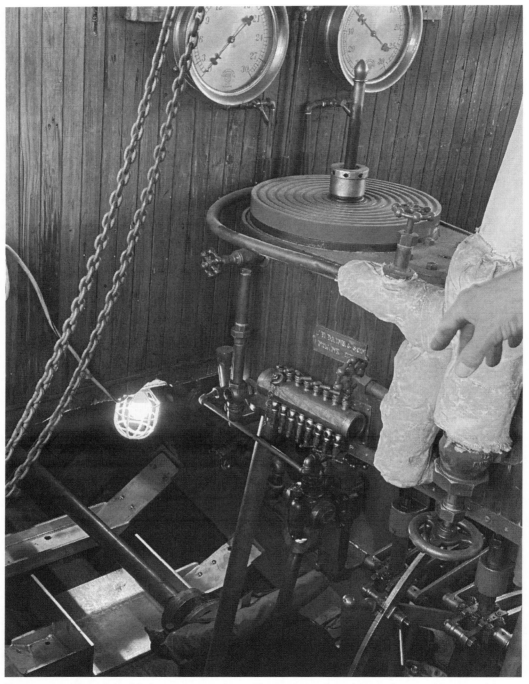

Figure 4.4.   The original Paine steam engine will remain in place. Mystic Seaport Museum, Mystic, Connecticut, 2022. Photo by author

for considering and comparing historic vessel technologies. Maintaining (and occasionally operating) the steam engine, which will be situated literally alongside the electric engine, allows visitors to see the options and the differences in energy use and technologies, and it brings *Sabino* into the twenty-first century and in line with environmental concerns. The museum could situate the accommodation into a larger sitewide narrative stressing the role of energy in maritime history, and reinforce the site's developing sustainability emphasis at the same time.

The theme of energy transition emphasizes part of the historical significance of the *Sabino* itself: throughout its life, it adapted to meet changing economic conditions. The Mystic Seaport interpretive staff already explores alternative narratives, and one about the evo-

Figure 4.5.   Whaling exhibit, Mystic Seaport Museum, Mystic, Connecticut, 2021. Photo by author

lution of energy resources could follow the scope of energy transitions illustrated through watercraft and transportation history, beginning with the historic whaling ship further down the pier. The sailors on the *Charles Morgan*, the last wooden whaleship that was part of an American whaling fleet, collected whale oil for the lighting industry. The museum is already adding signage to include more stories across diverse populations including the activities of African American and Native American people. Signage includes QR codes as well. The whaling exhibit around the *Charles Morgan* already uses the discussion to highlight the need for whale conservation. They could do something similar for energy in the interpretation of the *Sabino* to emphasize energy conservation, reduction, and transition.[29]

CASE STUDY: The Durango and Silverton Narrow Gauge Railroad, Durango, Colorado

Similar to steamboats, the steam train also shoulders tremendous historical and cultural importance. Historian Earl Pomeroy's classic *In Search of the Golden West* (1957) reviewed how nineteenth-century railroads brought wealthy Eastern travelers west, "peering through Pullman palace car windows at the remnants of the frontier."[30] These tourists scorned undeveloped Western lands and their Indigenous peoples, and so Western resorts and hotels like those run by Fred Harvey lured visitors in by replicating European cultural landscapes. Durango, Colorado, began offering scenic tours via train in July 1882. Later tourists then longed for the distinctiveness of the Western character. Early twentieth-century Durango boosters tried to tap the newer tourist market with historic and health attractions like Mesa Verde and the Trimble Hot Springs.

It was about the time of Pomeroy's publication in 1957 when gas and diesel began to dominate most transportation engine technology. Nostalgic railroad buffs and volunteers, primarily male, responded by promoting the steam train as part of a disappearing Western, even American, heritage. That same year, steam enthusiast and promoter Oliver Jensen wrote a lengthy ode in the popular magazine *American Heritage* entitled "Farewell to Steam," where he referenced many admirers: "They turn out by the thousands for one 'last' ride after another; they swap endless pictures, spikes, tickets, old timetables, even recordings of railroad noises; and they jabber away happily in professional jargon."[31] The idea of the train ride as a leisure activity arose from popular media and the toy and hobby industry between 1880 and World War II, but as planes and automobiles replaced the freighting and transportation purpose of trains, the railroad as a tourist ride gained immense popularity.[32] As the Cold War threatened cherished American values and comforts, the railroad as a tourist ride gained immense popularity. Between 1947 and 1962, ridership along the scenic Rio de las Animas Canyon rose from 3,400 passengers to 29,000, and locals wanted to tap the 800,000 visitors who came through Durango on their way to or from Mesa Verde and the Four Corners. The *Durango Herald* reported that most tourists fell into one of two groups: "Those who came to fish and those who came to gaze."[33]

Many predicted that heritage would enable long-term economic opportunities. The federal government helped rebrand the Four Corners region through the Area Redevelopment Administration (ARA). Created by Congress in 1961, ARA was part of a postwar response to New Deal programs that embraced Keynesian economics in hopes that private investment would stimulate economic growth at the local level through public works and public buildings.[34] The overextended and underfunded agency was short-lived, but the idea of tourism as a panacea for the postwar West gained enthusiasm. Cities like Santa Fe enlisted historic preservation to distinguish their Western identities. The regional office of the National Park Service (NPS) conducted studies between 1952 and 1962. One NPS report concluded that, while deserving of national recognition, the special skills involved in preserving and operating the train would be difficult for the agency to sustain.[35] Colorado banker William White proposed operating the railroad through his wife's Helen Thatcher White Foundation in December of 1959. White hoped to "invest in Durango as a tourist center" and turn Silverton into a "period piece."[36]

Priorities focused upon tourist experience and economic development over education and interpretation. Professor Robert E. Waugh, on leave from the University of Arizona, proposed that "Durango and the Rio Grande have a chance to be leaders in developing for the West attractions which may well rival the fabulous Williamsburg restoration for national attention."[37] Waugh predicted tourists would spend one billion dollars a year in the Four Corners region with attractions like Mesa Verde and Purgatory Ski Resort, which opened in 1960.[38] In 1962, businessmen hoped to "interest local people to operate concessions in the tradition of the early western railroad and mining days." They considered several

Durango properties from 4th Street to 6th Street "as ideally suited to this purpose" and hired the Walt Disney Corporation to survey Durango for year-round tourism potential.[39]

This newer manifestation of the Durango railroad founded a development company to restore the downtown it had platted over a half century earlier to serve as "Part II" of an Old West experience.[40] Attracting 11,611 passengers by July 4 (up from 8,778 the previous year), the *Herald* promoted the railroad ride with refurbished, private rail cars. An ad also ran: "The Year 1881, the visions of a Railroad to Durango came true, bringing growth and prosperity to the San Juan Basin. . . . Today, again, the Railroad's future vision promises to play an important part in the growth and increase of our areas."[41] Across the country, "railfan" club leaders chartered trains on little used tracks, boasting that they had "as many devotees as stamp collecting."[42] However, railfan publications promoted tourism in southwestern Colorado specifically by inviting riders to experience a time warp called *The Train to Yesterday*, arguing that "this historic little railroad must not be permitted to die because of a lack of friends or patronage, for it is as indigenous to Colorado as are the famous Cable-cars to San Francisco or the Statue of Liberty to New York."[43] Thus, Durango embraced tourism as annual ridership hit over one hundred thousand by the 1960s.[44]

Tourists arrived in Colorado eager to experience steam railroads and Western scenery with local businessmen urging them on, but railroad executives had no interest "in the entertainment business."[45] Trucks undermined the railroad freighting business, and the accessible roadways they used also encouraged more visitors. From 1951–1968, the extensive southwestern Colorado highway system compelled the railroad to discontinue several lines, first Durango to Alamosa, leaving railroading to tourism.[46] In the meantime, railroad management "never really understood the fan" who protested when the Denver and Rio Grande continued to shut down segments of its transportation services.[47]

While Amtrak struggled to fill its cars, riders continued to pay top dollar to experience this outdated mode of transportation and to view picturesque, scenic landscapes otherwise inaccessible to them. There were dozens of heritage railways by the end of the twentieth century, with nineteen in California, fifteen in Pennsylvania, eleven in Colorado, and around three more dozen in New England. Some of the most successful or at least well-known are in the West. They include the sixty-four-mile Cumbres and Toltec Narrow Gauge Railroad in New Mexico (the nation's longest and highest narrow-gauge steam railroad) and the "skunk" train in California, ironically fondly remembered for its stinky fumes. Through mergers and line closures, these tourist lines consolidated.[48]

With some exceptions, most heritage railways no longer serve the purposes of mere transport, but rather operate as for-profit historical experiences. Train enthusiast and founder of a tourist train along the Connecticut River, Oliver Jensen noted one Western train attraction with particular enthusiasm:

> Perhaps the most outstanding example of what this organized enthusiasm can do is the story of the narrow-gauge Silverton passenger service of the Denver & Rio Grande Western Railroad, last survivor of a network of narrow-gauge lines hacked out of the Colorado mountains many decades ago—last, indeed, of all narrow-gauge passenger lines in America. A few years ago, it had dwindled to a twice-weekly mixed train, with a single passenger car, and application was made for its abandonment. Then the steam admirers took notice

and moved in, until now, throughout the summer tourist months, the astonished railroad runs a train every day, with all its ten surviving cars packed solid. Not the least of the lures is that the power at the head end is honest old-fashioned steam.[49]

National papers including *The New York Times* similarly endorsed the Durango train repeatedly over the years, noting the authenticity of the experience. In 1999, Kathryn Jones wrote:

This was better than one of those simulated 3-D thrill rides at a theme park, and we weren't even going fast. The train travels an average of just over 13 miles an hour during the 45-mile trek from Durango up to the old mining town of Silverton. But the special effects were multisensory: the acrid smell of the train's smoke, the piercing whistle echoing

Figure 4.6.   Author in the centennial year of the D&SNGRR, 1981. Durango, Colorado. Photo by John S. Glaser

down the canyons and the rocking rhythm of wheels rolling across rails—not to mention the vertigo-inducing views.[50]

Travel writers and tourists repeatedly cite southern Colorado's train ride from Durango to the former silver mining town of Silverton, originally a part of the Denver and Rio Grande Railroad (the last private company to sell to Amtrak in 1982), as a must-do activity on anyone's itinerary. The roughly six-hour round-trip ride was one of the highlights of this author's own family "trip out west" in 1981 (the railroad's centennial birthday).

Echoing the iconic image of Western expansion, Durango's railroad, quite simply, defines the community's identity, yet it simultaneously undermines many of the residents' environmental sensibilities, health, and safety.[51] For decades, the Durango and Silverton Narrow Gauge Railroad (D&SNGRR) spewed black coal smoke into the air and often on riders in the open-air gondolas on its wildly popular and scenic train ride. The trip typically burns about six tons of coal to heat the water for steam, with multiple trains running during the summer season. This amount still pales in comparison to a coal-fired electrical power plant, like the one on the outskirts of Denver, that in 2003 burned close to 1,700 tons a day.[52]

As the train's coal-fire engine fueled the city's economic engine, it also smothered concerns about air pollution. Durango residents and train operators faced "profit and progress butting heads with quality of life . . . clean air became a worry, a concern, and a cause."[53] Beginning in the 1990s, South Durango residents noticed and objected to the black clouds that drifted toward their homes from engines running in the roundhouse at night to keep the engine boilers warm enough for morning departures. Richard Agee, who moved to the area about 1989, expressed the sentiment of many: "I love the train, but the smoke is a nuisance. . . . When I do remodel work, black soot simply POURS out of dead spaces in the walls. If it's getting in there, it's getting into the lungs of everyone who lives in South Durango, where most of it blows."[54] Another resident sarcastically resented the prioritizing of tourism over living. "Oh, and let's give the tourists what they want for ONE DAY while on vacation so they can go home to LIVE in their clean air, while Durango suffers for 365!" One acknowledged the issues of historic authenticity and heritage. "It's nice to want to give tourists 'perfect history' in order to make more money, but there is a reason that things change and it's unfair to make the Durango residents live in the unhealthy past so a private company can thrive."[55] In 1996, locals formed the South Durango Neighborhood Association, asking the Colorado Department of Health to investigate emissions from the trains.[56] Train operators cited their compliance with federal air quality regulations, which air pollution monitors supported, and insisted that the company maintained its engines, "the lifeblood of this company."[57]

In 1997, James Brooke of *The New York Times* observed the renewed but familiar dynamic between the railroad operators and residents, this time regarding rights to tourism's profits from photographs and obligations to pay for forest fires or pollution trains had caused. He interviewed a University of Colorado history professor Thomas J. Noel, who stated: "These small towns have always been at the mercy of the railroads. Today, it's just like the rail barons of old. You have a private meeting with the town, and you tell them what to do."[58] The upper hand came from the enormous economic benefits the railroad reaped, bringing in millions of tourism dollars. "If the train weren't here," replied Amos

Cordova, D&SNGRR vice president, "Durango wouldn't exist." And, he added, "The train is the only reason that Silverton exists."[59] When owner Charles Bradshaw sold the line to a Florida-based "entertainment rail company" in 1997, locals opposed building a "fun" train like the one that runs from Fort Lauderdale to Orlando. Allen Harper of American Heritage Railways then purchased the lines, promising to pursue steps to reduce smoke emissions at the roundhouse.[60]

In 2001, Harper installed two scrubbers at a cost of $400,000 to collect particulates from idling locomotives. Train emission levels fell well below EPA standards, but complaints rose again in 2005. The EPA's monitoring station for emissions supported those assertions. To address issues like indoor air quality and asthma, Colorado's Public Health Department considered regulating the state's sixteen other historic steam engines with stricter guidelines than those of the Clean Air Act, which exempted locomotives built before 1973.[61] The railways were not impressed. "If the state wants to regulate us, they should be prepared to pay the bill," said Leo Schmitz, director of the Antonito-based Cumbres & Toltec Scenic Railroad Commission."[62] Wasatch Railroad Contractors (WRC) of Cheyenne, Wyoming, evaluated the possible use of alternative fuels for overnight storage and proposed better handling of the ash piles, repairs to the smokeboxes and draft and combustion systems, procedures for overnight storage of the locomotives while at the Durango Roundhouse, changing crew procedures, and adopting several alternative fuels for overnight use, such as natural gas (and other alternative liquids) and biomass (wood pellets).[63]

Harper adopted several recommendations. These included building an ash pit in Silverton and developing new procedures, such as using diesel engines to keep engines warm at night rather than coal-fire steam. Task force members attested that while Harper was "committed to reducing smoke emissions at the roundhouse, he's also fiercely committed to preserving the historic nature of the locomotive." "This is not an amusement ride," said Harper. "If you want to ride an electric-powered train, go to Disneyland." Citing a commitment to historical accuracy, he and ardent supporters steadfastly refused to abandon coal as a power source.[64]

By 2010, Harper pledged to spend $1 million over five years to reduce emissions by 10 percent annually. The D&SNGRR adopted many of the WRC feasibility report's recommendations including: burning wood pellets, using diesel instead of coal-fire steam locomotives for switching trains and performing track maintenance at the Durango station, giving all engine firemen specialized training on how to place coal and wood pellets into the coal box to facilitate cleaner burning, and purchasing green power to offset greenhouse-gas emissions from the railroad's vehicle fleet by planting 2,587 trees in Durango and Silverton.[65] Harper claimed to have "erased" the railroad's carbon footprint: "We are the only entirely 'green' railroad in the world."[66]

Unfortunately, air quality was not the only environmental risk that the coal-fire engine posed. Steam trains historically caused fires, and they served as an impetus for forest conservation as spark arrestors caused repeated fires.[67] Droughts exacerbate burn risk and climate change has prolonged droughts. On the morning of June 1, 2018, hot cinders ignited a drought-starved landscape, causing the sixth-largest wildfire in the state's history, burning 54,000 acres across southwestern Colorado until July—shutting down the San Juan National Forest and necessitating the evacuation of thousands. The railroad suspended

Figure 4.7.   Durango-Silverton Narrow Gauge Railroad, Durango, Colorado, 2011. Photo by author

operations for forty-one days, costing the region millions in revenue. Once a federal investigation proved the cause, two dozen area businesses, residents, and the U.S. Attorney's
Office sued for the $25 million spent on the fire. With a heavy economic dependence on
the railroad, not everyone agreed with the lawsuit. Still, the strain urged Harper to make
changes a year later, agreeing to spend $6 million for two diesel engines to use on days of
heightened risk.[68] "In an era of growing concern over man caused climate change, blasting
6 tons of burnt coal emissions into the clean mountain air of Southwest Colorado seems
to have gone unnoticed."[69] The "416 fire" had been just the latest of over four dozen fires
the railroad caused that year. Finally, the consumption of six tons of coal and ten thousand
gallons of water per trip, not to mention the coal emissions, finally helped turn the tide of
support toward environmental conservation.

In early 2020, fire and air monitors confirmed longtime complaints about air quality. An
advocacy group formed called "Sustain the Train" to call for both safety and economically
sustainable operation. The Durango & Silverton Narrow Gauge Railroad announced that it
was converting to oil. Harper admitted that "about eighty percent of riders don't come for
a coal-fired locomotive; they come for the steam-engine *experience*, which oil will provide."
Harper identified a 1902 locomotive that had been used for display as a candidate for conversion to oil, which, while still a fossil fuel, does not have the same levels of fire and air
quality risks. Passengers will still observe steam escape out of the stack and hear the whistle.
The roundhouse would still maintain coal-fired engines as a presence.[70]

Artifact Spotlight

The Coal-Fired Steam Engine

The heritage train's rise in popularity through the 1970s barreled headlong into the environmental movement focused upon clean water and air, and eventually the climate crisis. The continued popularity of historic scenic railways across the globe must provoke public historians to think critically about our attitudes toward energy technology and its relationship to the environment, in both the past and the present. How should preservation professionals balance nostalgia, tourist dollars, historical accuracy, and sense of place with environmental sustainability? The railroad is vital to understanding economic development and western expansion, but it is a particularly problematic historical resource to interpret and preserve from an environmental standpoint. The continued popularity of these experiences provides us the opportunity to prompt a wider audience to think more critically about our historic and contested attitudes toward technology and its relationship to the environment.

While one experience may be obviously "fake," the heritage railways share with Disneyland the goals of fun, relaxation, and entertainment. They both tap into an almost religious emotion that Catherine Cameron and John Gatewood refer to as *Numen*. Numinous sites can increase visitor excitement and enthusiasm.[71] Hal Rothman wrote about leading senior citizens on a train tour of the national parks in 1999 where the train was refurbished as a 1940s luxury car, giving customers a not-quite-realistic journey into the past. But "cultural tourism has come to mean a great deal more than most scholars envision, education about the past," he admitted. "It has become an industry that brings together a combination of history, nostalgia, myth, and entertainment, in the guise of teaching about American Culture."[72] With some exceptions, the driving philosophy guiding the interpretation and preservation of the resource is to provide tourists with a numinous feeling of having an "historical experience." Such experiences privilege the tourists (and hence economic development).

Heritage railways are different from almost any other type of cultural resource because, for the most part, cultural resource managers do not manage them. Like the original railroads, corporations do. One hopes and assumes that these managers do not fall to the depths of the robber barons Richard White describes in his 2011 book, *Railroaded: The Transcontinentals and the Making of Modern America*. While the operators may believe firmly in historical integrity, the second motive is profit. Al Harper defended this:

> The best way to preserve history is to make the experience so exciting it will pay for itself. Some people want to preserve history just for the sake of history and not to develop things that will bring in those bigger crowds, and you end up with subsidized history that may require government support for a long time.[73]

There is nothing wrong with celebrating old technology, especially one so integral to larger historical processes and contexts. But likewise, there is benefit in celebrating new technologies that respond to today's needs. Gunter Dinhobl articulated the conundrum of preserving energy technology for "rolling stock" in particular. He contended that true preservation would mean maintaining a "dead" technology. Use requires adaptation and repair.[74] Stewards of historic properties must make tough decisions when technologies have either worn out or become obsolete due to environmental standards, energy efficiency, or technological advances. Replacement and mechanical upgrades can become part of the interpretive story, with original pieces remaining "in place," if possible, to illustrate transition and change. (See figure 4.4.)

Back in his 1995 publication, *Mickey Mouse History*, historian Mike Wallace called upon public historians to offer alternatives to commercialized history, boldly claiming that, "The past is too important to be left to the private sector."[75] Rather than try and compete, historians can help redefine the significant and relevant issues by working with private companies. As an example, the railroad's attempt to distinguish between history and entertainment is not entirely accurate. Entertainment and economic development promoted the rehabilitation of the train largely for purposes of entertainment, not historical education. The revival of Durango as a tourist center around its railroad was a deliberate economic decision that evolved out of the successful development of destinations like Disneyland, which opened in 1955. Disneyland's opening-day film footage shows a re-creation of a Santa Fe Railroad engine with Walt Disney himself at the controls alongside the actual Santa Fe Railroad president, the California governor Goodwin Knight, and Mickey Mouse (whose origins are tied to another industrial classic as Steamboat Willie), circling the park and tooting the whistle for a mile and a half around a six-minute track that marks the clear boundary of Disneyland. The train engine duplicated those from over fifty years prior (but at five-eighths the size) to support the park's mission "to relive fond memories of the past . . . dedicated to the ideals, dreams, and hard facts that have created America with the hope that they will be a source of joy and inspiration to all the World." Like the railfans advertised about the Durango-Silverton experience, tourists could "travel through the gates of time."[76] That same train exists today as well as the Big Thunder Mountain Railroad, which hurtles guests through the dangerous and abandoned landscape of the mining world. When they are done, like at Silverton, one can shop and dine at various establishments and soak in the atmosphere of Frontierland.

Smoke as Material Culture

The dispute over smoke reveals more about the tense partnership between cultural tourism and historical interpretation. Attempting to alleviate air pollution is not ahistorical and can engage tourists in railroad history beyond the riding experience. Again, celebrating old technology, especially one so integral to larger historical processes and contexts, must not mean rejecting new technologies that respond to today's contexts. Fuel transitions have not been linear. The first locomotives on the Transcontinental Railroad fired wood, but with the abundance of coal, railroads allowed the use of different types of fuel. At Promontory Point, where the Transcontinental Railroad was completed, one locomotive was wood-fired Jupiter ("designed to arrest sparks and prevent fires") and the other, No. 119, burned coal.[77]

While known for the pristine beauty of the desert plateau, the Four Corners are also home to multiple sources of energy extraction, air pollution, cultural landscape desecration, and ecological destruction, particularly through private development on the Indian reservations. The Durango solution addresses the problem of pollution, but it does not address the problem of how the public engages with industrial history. The unquestioning acceptance that soot and smoke are necessary for understanding the history of steam power, or even economic prosperity, invites introspection. True, the smoke from the tourist train is hardly the region's worst environmental offender, but with a captive audience, already fascinated by nineteenth-century technology, smoke from industrial centers, power plants, steamships, and trains can be an extremely effective window for educating the public about how twenty-first-century technology can address the problems that people wanted to, but could not, fix over a century ago.[78]

Historians can certainly analyze smoke in its gaseous state as material culture. It has several properties associated not only with touch, but sight, smell, and taste, as well. Steam preservationists cite these very qualities. How people have responded to those qualities also has a relevant history for steam technology. The public has always disputed smoke's necessity and value and interpreting that debate historically can help visitors understand today's preservation decisions. In a May 1895 magazine article accompanied by Thomas Moran illustrations, the writer argued that smoke actually detracted from a passenger's experience:

> In the front rank of the great railway systems of the world, it seems probable that the Denver and Rio Grande, owing to the topography of the country through which it passes, will be one of the first to be converted from the old ways of steam to the new world of electricity. Along its devious routes, a hundred mountain streams waste their energies. Down every mountain-side dash waters capable of bringing dynamos of countless horse-power. . . . First to welcome the new electricity will, I predict, be the mountain railway system of the Rio Grande, and the transcontinental traveler will, in the near future, enjoy its magnificent mountain views free from the nuisance of smoke and cinders.[79]

As this sentiment implied, most residents and train operators during the heyday of this transportation technology would have supported the technological changes that reduce smoke emissions. Like today, passengers and downwind neighbors during the Progressive Era *were* complaining of railroads over the "smoke nuisance."

Complaints about train smoke for environmental, aesthetic, and health reasons are not a twenty-first-century phenomenon. Local antismoke legislation across the country goes back to the Progressive/Conservation Era alongside the golden era of the nation's trains. Historian David Stradling has documented the "the growth of an environmentalist sentiment in turn-of-the-century cities."[80] John Stilgoe writes that even engineers came to view smoke as evidence of wasting fuel. Dark smoke meant incomplete combustion, and inspectors cited the firemen if the smoke was too black. Smoke stoking equipment helped. Fuel efficiency and cost savings also supported the smoke abatement movement.[81] In November 1907 at the American Civic Association in Providence, George W. Weldon of the New York, New Haven, and Hartford Railroad announced that they were interested in both cost savings and the comfort of patrons. He claimed that the New Haven Railroad would be more than willing to adopt technology that could completely burn the gases that cause the black smoke without much concern for its cost.[82] Years later in 1954, the *American Railway Times* published a story: "Railway Travel sans Dust, sans Smoke, sans cinders, sans misery," excitedly reporting about the coal-fired steam engine associated with the Albany and Buffalo

Lightning Express coach, which had added tanks of water with thick, coarse sponges to filter the dusty air before it gets into the passenger car.[83] Technological change and adaptation, when carefully applied, can tell stories about change and evolving needs.

The popular understanding of historic preservation is problematic for historic but operational steam vehicles because, to the general public, preservation implies rigid stasis. Railfans and train owners might argue that any modification will destroy the historic integrity of the technology. However, the secretary of the interior's standards allows updates to historic homes and buildings to meet energy efficiency standards as long as one can distinguish historical materials from new and the overall "feeling" and "association" survive the adaptations. Several of these railways, like the D&SNGRR and Cumbres and Toltec, made other physical modifications to their trains, having replaced or "re-created" wooden passenger cars with custom-built coaches, added lighting, or expanded seating capacity.[84] Yet seemingly sensible compromises to reduce environmental impact have drawn accusations that successful entrepreneurs have submitted to environmentalists' political agendas. While the Durango and Mystic solutions might not have accommodated the views of the purists, both historical interpretation and preservation need flexibility to contribute to present-day narratives.

Interpretation is about the message, and perception is a large part of that message. As Leo Marx attempted to do, we need to contextualize and reconcile "the machine in the garden" when we interpret historical energy use. Educators and interpreters can engage visitors in discussions about the relationship between technology and the environment, both in the present and historically. A historic resource needs historic interpretation that resonates with contemporary audiences. The past has not passed. The dispute over smoke reveals the tense, complicated, and often opposing partnership between cultural tourism and historical interpretation and preservation. We are not disconnected from these concerns. "Grandfathering" in historic energy technology in the name of authenticity is not a responsible choice.

Both the D&SNGRR and Silverton are National Historical Landmarks, designated as such in 1961, prior to the National Historic Preservation Act of 1966 and the creation of the National Register and the Secretary of the Interior's Standards for the Treatment of Historic Properties. This landmark status means the National Park Service has determined that these resources met a different standard of historical significance than a mere National Register listing. Air quality is a primary concern for Durango heritage organizations, prompting many to adopt sustainability as part of their mission to protect the decay of historical resources like nearby Mesa Verde and Chaco Canyon, not to mention the toll that emissions and particles take on historic buildings in places like Durango's Main Street.

Historians define *history* as "change over time," and thus change can and should be part of historical interpretation. This makes sense for historic energy resources, but past stewardship has not adopted this philosophy. *Sabino* claimed historic designation to bypass the environmental regulations that prohibited operating coal-fired engines with true "authenticity." However, the reasoning that historic energy resources or sites of energy development will lose their historic significance if they no longer operate or use the same historic energy resource is flawed. Historic preservation does not necessarily mean freezing a resource, but rather honoring it to the extent that it reflects a period of significance. Integrity is determined by significance. Even buildings evolve to respond to changing use and technology, and historic resources lose even more significance if they are no longer used. While the National Register designates properties for recognition, it does not require they be frozen in time. Section 106 of the National Historic Preservation Act authorizes many different ways to preserve a resource, which often necessitates the replacement of technology. The secretary of the interior's standards allow owners to update historic homes and buildings to meet energy efficiency standards as long as the overall "feeling" and "association" survive the adaptations. One way to mitigate the loss of historic technology is the detailed documentation required in a HAER (Historic American Engineering Record) report.

The prevailing historic preservation policy since the National Park Service began stewarding historic properties has remained fairly constant—to make the smallest impact possible and to continue to save the resource for the history it conveys: "Better to preserve than to repair, better to repair than to restore, better to restore than replace/reconstruct."[85] Significant features may eventually lose out to changing needs and modern necessity; however, the smallest impact on the significant features is usually preferred. In the case of *Sabino*, while the coal-fired steam engine may no longer always be in use, it still retains other character-defining features of small steamboats, and the engine and coal remain important parts of the resource's significance. Mystic has decided not to remove the engine to preserve it for the public and future generations. Additionally, the steam engine is often only just one piece of what makes a resource significant. Decisions about preservation are "not always the year the exhibit was built, because structures evolve to meet changing needs," and those needs may indeed be historically significant.[86]

Following a strategy of "preserve in place" allows the original technology to remain in its original setting. At the same time, for practical concerns, resource stewards are often able to maintain these mechanicals in working order in case the newer technology, often dependent on a computer or digital technology, fails. This makes the final decisions regarding the *Sabino* and the Durango Narrow Gauge Railroad examples of best practices that not only preserve the historic technology, but they address the contemporary concerns about climate change. As for steam technology, we should preserve it for demonstrations, but instead of using fossil fuels, renewable energy such as a solar or an electric battery might generate power for excursion experiences.[87]

With today's need for sustainable solutions, the Mystic Seaport Museum's steamboat and Durango's narrow-gauge train offer lessons about resolving conflicts over historic integrity and energy use through the shared authority of all stakeholders. The high profile of the train and the steamboat as cultural and technological symbols, often tied to local identity, necessitates dialogue. In addition to facilitating these discussions, public historians should help address public misconceptions and attitudes about technology and environmental issues in the past. The steam engine had great importance to America's industrial history, but its narrative can describe and emphasize evolving energy technology and reasons for energy transition. Museums and historic site interpretation should facilitate discussions that address public misconceptions about historical authenticity and attitudes toward environmental issues in the past as well as the present. Discussing and highlighting those compromises offers even more opportunity for interpreting energy.

Notes

1. Louis C. Hunter, *Steamboats on the Western Rivers: An Economic and Technological History* (Cambridge, MA: Harvard University Press, 1949); Henry Dickinson, *A Short History of the Steam Engine* (London: Macmillan, 1939); Carroll Pursell, *Early Stationary Steam Engines in America: A Study in the Migration of Technology* (Washington, DC: Smithsonian Institution Press, 1969); R. Douglas Hurt, *American Farm Tools: From Hand-Power to Steam-Power* (Lawrence, KS: Sunflower University Press, 1982).

2. Richard L. Hills, *Power from Steam: A History of the Stationary Steam Engine* (Cambridge: Cambridge University Press, 1993); Maury Klein, *The Power Makers: Steam, Electricity, and the Men Who Invented Modern America* (New York: Bloomsbury Press, 2008); William Rosen, *The Most Powerful Idea in the World: A Story of Steam, Industry, and Invention* (Chicago: University of Chicago Press, 2010).

3. Andreas Malm, *Fossil Capital: The Rise of Steam Power and the Roots of Global Warming* (Brooklyn, NY: Verso Books, 2016).

4. Many other similar museums hold events like this, including the Connecticut Antique Machinery Association in Kent, Connecticut, and the Powerland Heritage Park in Oregon.

5. Mystic Seaport Museum data sheet, "Vital Statistics" (Mystic, CT: Mystic Seaport Museum, n.d.), Mystic Seaport Library, Mystic, CT.

6. George King III, *A Steamboat Named Sabino* (Mystic, CT: Mystic Seaport Museum, 1999), 76–79.

7. Registrar files, Mystic Seaport Library, Mystic, CT.

8. L. Revell Carr to Waldo CM Johnston, Memorandum, "The Master Plan for Mystic Seaport," March 13, 1973, Mystic Seaport Library, Mystic, CT.

9. J. I. Little, "Scenic Tourism on the Northeastern Borderland: Lake Memphremagog's Steamboat Excursions and Resort Hotels, 1850–1900," *Journal of Historical Geography* 35 (2009), 716–17.

10. Sarah Willis, "Sabino: Yesterday and Today," 4–5 (Maritime History, Professor Labaree, November 16, 1978), RF 336, Mystic Seaport Library, Mystic, CT.

11. King, 84.

12. Mystic Seaport Museum data sheet, "Vital Statistics" (Mystic, CT: Mystic Seaport Museum, n.d.); Nicholas Dean, "Sabino," National Register of Historic Places Registration Form, National Park Service, United State Department of the Interior, September 1991.

13. King, 84.

14. Robert W. Morse, "The Restoration/Reconstruction of the Steamboat Sabino. . ." unpaginated typescript, 5 (RF 336, Connecticut Mystic Seaport Library, Mystic, c.1980) (Morse Report).

15. King, 103–05; Willis, 13.

16. King, 76–79; Carr to Johnston.

17. Willis, 7–8

18. Willis, 7–8.

19. Willis, 8, 13.

20. Willis, 11; A. C. Hardy, *American Ship Types: A Review of the Work, Characteristics, and Construction of Ship Types Peculiar to the Waters of North American Continent* (New York: D. Van Nostrand Company, Inc., 1927), 96.

21. King, 82; Detailed logbooks, Box 12/4 1975; Box 12/6 1977, Mystic Seaport Library, Mystic, CT.

22. Willis, 6.
23. Willis, 11.
24. Richard Lotz, "SS Sabino: Still Steaming," *Seaways* (March/April 1991) (Master of SS *Sabino*, 1979–present), 16.
25. Email correspondence, Mark Potash to Leah Glaser, June 3, 2013.
26. Kate O'Mara, "Modestly Serving the Community," https://educators.mysticseaport.org/artifacts/sabino, accessed May 23, 2021; "Worksheet 1: Perspectives: Steamship Sabino: Modestly Serving the Community, Mystic Seaport: The Museum of America and the Sea," https://educators.mysticseaport.org/static/connections/pdfs/sabino_worksheet_1_perspectives.pdf.
27. Maria Gallucci, "This 110-year-old Steamboat is a Floating History Lesson," November 2, 2018, https://www.atlasobscura.com/articles/sabino-steamboat; "Steaming Again," Mystic Seaport, July 21, 2017, https://www.mysticseaport.org/news/steaming-again; "Steamboat Sabino in Mystic May be Going Green," NBC Connecticut, January 2, 2020, https://www.nbcconnecticut.com/news/local/steamboat-sabino-in-mystic-may-be-going-green/2204820; Joe Wojtas, "Mystic Seaport Looks to Convert Steamboat Sabino to Electric Power," *The Day*, January 1, 2020, https://www.theday.com/article/20200101/NWS01/200109940.
28. Chris Gasiorek, September 3, 2021, "Uncontrollable lithium battery fires at sea," http://www.lithiumsafe.com/battery-fire-safety-marine, accessed September 3, 2021.
29. Zoom meeting with Leah S. Glaser and Chris Gasiorek, March 10, 2021.
30. Earl Pomeroy, *In Search of the Golden West: The Tourist in Western America*, reprint (Lincoln: University of Nebraska Press, 1990), 7.
31. Oliver Jensen, "Farewell to Steam," *American Heritage* 9: 1, 1957, https://www.americanheritage.com/farewell-steam.
32. John R. Stilgoe, *Metropolitan Corridor: Railroads and the American Scene* (New Haven, CT: Yale University Press, 1983), 4.
33. "White, Tutt, Play Major Roles," *Durango Herald*, January 10, 1963; "Most visitors come to fish, some just look," *Durango Herald*, June 8, 1947.
34. Gregory S. Wilson, *Communities Left Behind: The Area Redevelopment Administration, 1945–1965* (Knoxville: University of Tennessee Press, 2009).
35. "Durango-Silverton Narrow Gauge Railroad: A Study" (Region 2 Office, National Park Service, Department of the Interior (March 1962), 52.
36. Duane A. Smith and Elizabeth A. Green, *Seasons of the Narrow Gauge: A Year in the Life of the Durango & Silverton* (Durango, CO: Durango Herald Small Press, 2011), 14.
37. "Four Corners Tourist Mecca Urged by Waugh," *Durango Herald*, January 11, 1963.
38. Gene Wortsman, "Four Corners Tourist Mecca Urged by Waugh" (January 11, 1963). Waugh consulted for the BIA soon afterward.
39. "D&RGW Makes Plans to Develop Millions of Dollars' Worth of Property" and "Railroad to Restore Buildings," *Durango Herald* (January 10, 1963).
40. Alexis McKinney, manager of the Silverton branch, explained that "it is a two-part project . . . as a permanent passenger-carrying railroad, embodying as authentically as possible the appearance, service, and traditions of early railroading in the Rocky Mountains, and second to develop the 400 and 500 blocks of Main Avenue as tourist attractions reminiscent of the Old West." "How the Rio Grande Land Was Born," *Durango Herald* (April 2, 1963).
41. "Two Trains to Silverton," *Durango Herald* (July 3, 1963).
42. "Railroad Fans," *Popular Mechanics* (August 1947), 80–87.

43. Thomas T. Taber, "Train to Yesterday," Durango Silverton Narrow Gauge Railroading in Southwestern Colorado (The Railroadians of America, 1955), Foreword.
44. Smith and Green, *Seasons of the Narrow Gauge*, 16; Duane Smith, *Rocky Mountain Boom Town: A History of Durango, Colorado* (Boulder: University Press of Colorado, 1992), 44, 88, 166–67.
45. Smith and Green, *Seasons of the Narrow Gauge*, 9, 13.
46. Smith, *Rocky Mountain Boom Town*, 8–9.
47. Terry Berger and Reid, Robert, *Great American Scenic Railroads* (New York: E.P. Dutton, Inc., 1985).
48. Smith, *Rocky Mountain Boom Town*, 205, 233; Smith, *Seasons of the Narrow Gauge*, 16–17.
49. Oliver Jensen, "Farewell to Steam," *American Heritage* 9:1 (1957), https://www.americanheritage.com/farewell-steam, accessed September 26, 2002.
50. Kathryn Jones, "Three Trains Years Behind Schedule: Colorado; Durango to Silverton is a real cliffhanger," *New York Times*, August 29, 1999; "Scenic Tourist Trains Return to the Rockies," *New York Times*, 1999.
51. For more on this tension, see Hal K. Rothman, *Devil's Bargains: Tourism in the Twentieth-Century American West* (Lawrence: University Press of Kansas, 2000).
52. There are also numerous problems that make a far greater negative impact on the environment, including old uranium tailings, which the EPA removed from the banks of the Animas River in 1991; Kim McGuire, "Steam trains may have to clean up act. State regulators are zeroing in on coal-fired locomotives' trail of soot, ash and sulfur, but train operators are crying foul. [Final Edition]," *Denver Post*, July 9, 2006, sec. News.
53. Duane Smith, *Rocky Mountain Heartland: Colorado, Montana, and Wyoming in the Twentieth Century* (Tucson: University of Arizona Press, 2008), 240–41.
54. Associated Press, "Neighbors Blow Whistle on Train Soot," *Denver Post* (August 1997).
55. Woodsmokehaz1, "New Scrubber System for (antique) Train would Cut emissions {sic}," (comments), http://woodsmokehaz1.wordpress.com/2010/09/09/2010-sept-9-co-durango-comments-on-coal-smoke-new-scrubber-system-for-antique-train-would-cut-emissons-says-some-day-perhaps-clothes-i-put-on-the-line-wont-smell-like-coal-smoke-when-i-bri (September 09, 2010).
56. Durango Downtown.com, "Train Smoke Mitigation Task Force Effort Receives Funding" (August 27, 2009), https://durangodowntown.com/news/durango-and-around/train-smoke-mitigation-task-force-effort-receives-funding, accessed September 26, 2021.
57. Associated Press, "Neighbors Blow Whistle on Train Soot," *Denver Post* (August 1997).
58. James Brooke, "As Tourism Rides Rails, New Barons Gain Clout," *New York Times* (September 2, 1997).
59. Brooke.
60. Smith and Green, *Seasons of the Narrow Gauge*, 18.
61. Missy Votel, "Cutting through the haze: Local Health Department Offers Options for Pollution Reform," *Durango Telegraph* (September 15, 2005).
62. Kim McGuire, "Steam Trains May Have to Clean Up Act."
63. Members included the City of Durango and the Town of Silverton, La Plata County, South Durango residents, the San Juan Basin Health Department, Durango & Silverton Narrow Gauge Railroad, and representatives from Senator Michael Bennet, and Representative John Salazar's Offices. *Wasatch Railroad Contractors, Durango and Silverton Narrow Gauge Smoke Mitigation Feasibility Report* (Durango, CO: Region 9 Economic Development

District, October 2006), http://wrrc.us/about/highlighted-projects/durango-silverton-smoke -mitigation.

64. Kim McGuire, "Steam Trains May Have to Clean Up Act."

65. In 2009, the Economic Development Administration, the City of Durango, La Plata County, and the railroad itself contributed funds to install an energy-efficient scrubber system for each train smokestack; "Train Smoke Mitigation Task Force Effort Receives Funding," (August 27, 2009), www.downtowndurango.com; "Smoke Signals, D&SNG Continues Efforts to Clean UP Emissions," *Durango Telegraph* (September 3, 2009).

66. Smith and Green, *Seasons of the Narrow Gauge*, 135.

67. Mary Elizabeth McCahon et al., *Department of Environmental Protection Cultural Resource Survey, 1985* (Hartford: Connecticut Historical Commission, State Historic Preservation Office, 1986), 1; 2 April 1909–July 1912, Forester's office. Old letters, folder 1.

68. Sam Tabachnik, "Durango residents divided as beloved coal fired train faces lawsuits over its role in the 416 fire: A Federal Investigation Concluded that Hot Cinders from the Train's Smokestack Sparked the 2018 Wildfire," *Denver Post*, July 7, 2019, https://www.denverpost .com/2019/07/07/durango-silverton-railroad-lawsuit-fire.

69. Durangodowntown.com, "Is the Durango Train Worth the Environmental Impact in 2020?" https://durangodowntown.com/news/is-the-durango-train-worth-the-environmental -impact-in-2020, November 5, 2019, accessed September 26, 2021.

70. Jonathan Romeo, "Coal-fired Durango and Silverton Railroad Converting to Oil," *Denver Post* (February 24, 2020), https://www.denverpost.com/2020/02/24/durango-silverton -narrow-gauge-railroad-coal-oil.

71. Catherine Cameron and John Gatewood. "Excursions into the Un-remembered Past: What People Want from Visits to Historical Sites," *Public Historian* 22:3 (Summer 2000), 109, 127.

72. Hal Rothman, "Cultural Tourism and Changing Society," in *Public History and the Environment* (Malabar, FL: Krieger Publishing Company, 2004), 73.

73. Jason Blevins, "Single Manager to Run Cumbres & Toltec Railroad," *Denver Post*, October 5, 2012.

74. James Williams, "Conference Report: Reusing the Industrial Past," *Technology and Culture* 52:2 (April 2011), 377.

75. Wallace, 155.

76. 1955 Disneyland Opening Day [Complete ABC Broadcast], http://www.youtube.com/ watch?v=JuzrZET-3Ew, posted July 17, 2011, accessed September 26, 2021.

77. Mark Fiege, *The Republic of Nature: An Environmental History of the United States* (Seattle: University of Washington Press, 2013), 229, 245.

78. National Park Service, "Durango-Silverton: A Study," March 1962.

79. There is much to unpack from this source, like how it goes from describing running water as "wasted energy" to advocating for another energy system for trains. J. B. Walker, "Great Railway Systems of the United States," *Cosmopolitan: A Monthly Illustrated Magazine* (May 1895), 28.

80. David Stradling, *Smokestacks and Progressives: Environmentalists, Engineers, and Air Quality in America, 1881–1951* (Baltimore, MD: Johns Hopkins University Press, 2002), 2.

81. Charles H. Benjamin, "Smoke Abatement in Large Cities," *Outlook* 70, no. 8 (February 22, 1902): 480; Stilgoe, 121–22.

82. George Welden, "The Smoke Nuisance on Locomotives," *The Smoke Nuisance* in American Civic Association, Series II, No 1(March 1908; Second Edition, 1911), 31.

83. "Railway Travel sans Dust, sans Smoke, cans cinders, sans misery," *American Railway Times*, October 12, 1854, 6, 41.

84. Jason Blevins "New Way Ahead," *Denver Post* (October 5, 2012); "Sparkling, Fresh Engines for Narrow Gauge," *Durango Herald*, April 25, 1963.

85. As quoted in Barry Mackintosh, "The Case Against Reconstruction," *CRM Bulletin* 13, no.1 (1990).

86. Blevins "New Way Ahead"; King, 84.

87. Robert Green, *The Surprising Future of Steam Power,* TedxMissionViejo, TedxTalks, January 4, 2017, https://www.youtube.com/watch?v=SbgwDk7A4as&t=242s, accessed March 10, 2021; also see discussion on how the Collins Axe Company preserves its powerhouse without burning fossil fuels. Ken Byron, "Connecticut Town Finds What's Old Is New with Hydropower," manuscript, October 15, 2019, https://static1.squarespace.com/static/5f10f97dbd51f52f7c fae6c4/t/5f285b7d72017d631ac9b2dd/1596480381455/Canton+Hydropower.pdf.

CHAPTER 5

The Power of Fossil Fuels

Energy from Coal, Oil, and Gas

WHILE WATERPOWER AND STEAM TRAINS tend to occupy the most real estate in the historical imagination of the American industrial landscape, the extraction and burning of coal and other fossil fuels like oil and gas for manufacturing, motor vehicles, airplanes, and electricity assume the most responsibility for molding everyday modern society. In doing so, energy systems are so intertwined that fossil fuels easily replaced the heating and manufacturing energy systems designed for organic sources (namely wood and water) to meet the demands of manufacturing and industrialism.

Fossil fuels are all naturally occurring organic matter, like marine animals, plants, algae, and bacteria that have decayed over millions of years and formed liquids (oil), solids (coal), and gases (natural gas). Because of that process, however, they are finite resources. At some point in time, they will no longer be available for extraction, conversion, and use (at least not in terms of meeting the demand). These carbon-based resources burn very efficiently and produce heat, light, and gas. Anthracite (aka "stone") coal has a higher percentage of carbon, burning hotter and with less soot than bituminous, semi-bituminous, or lignite coal, but that also means it has few combustible gasses and requires special technologies to ignite. Burning releases carbon into the atmosphere in the form of carbon dioxide gas, which traps heat energy that would otherwise leave the earth. Thus, too much fossil fuel burning alters the atmospheric temperature, causing climate change.

Fossil fuels include secondary energy sources created through processing and refining. Coke is a hard, dark gray, porous fuel source. Its relationship to coal is similar to that of charcoal to wood. One produces coke by heating coal with limited oxygen, and it likewise burns hotter and cleaner than regular coal. Steam is the gaseous phase of water that occurs after heating the liquid to its boiling point, such as when burning coal to heat a boiler, or even converting solar (in solar thermal plants). High-pressure steam turns pistons to achieve motion. Connected to a generator, pressure from the steam spins a turbine. The spinning turbine then transforms the kinetic energy of steam into the energy of moving electrons to create electrical current, and this is carried along a conductor like copper wire.

The most well-known fossil fuel, petroleum (aka crude oil), can often be found in shale (sedimentary rock beneath the earth's surface), the former locations of ancient seabeds (sedimentary basins), and tar pits. Pressure and heat changed some of this carbon- and hydrogen-rich material into coal, some into oil (petroleum), and some into natural gas. American companies locate petroleum in reservoirs, often on the Gulf Coast or across the American West. In the United States, Americans mine coal in places like West Virginia, Pennsylvania, and the Colorado Plateau, and extract natural gas in Pennsylvania and the Gulf Coast.

Organic forms of energy like water or wood required far more human muscle in the past, which naturally limited their use for energy. Recall the difficulty that rural industrial sites in the Northeast faced when the source of their energy, a river powering a waterwheel, froze during the winter months. As popular as moving waterpower was, it was also not an option for more arid regions. Fossil fuels, on the other hand, offered utilities reliability and control while removing the drudgery and work of energy production from the users' experience. Utilities sold that convenience, and invisibility, to customers, and that access has become ubiquitous. At the same time fossil fuels began to infiltrate the natural and cultural environment, long-distance infrastructure in the form of transmission lines and pipelines hid the process of power generation and delivery. In the cases of oil and gas, the majority of energy users in high population centers became both intellectually and emotionally distanced from sites of production.

The fossil fuel industry has allowed consumers to use and expect access to energy at any time of day, in any region, and in any weather. Both cultural and economic dependency on fossil fuels for an energy-rich economy makes voluntary energy transition seemingly unrealistic. If so, what obligations do public-facing historians have to encourage the public toward a postcarbon, lower-energy future?[1] Senior fellow at the Post Carbon Institute, Richard Heinberg is not a historian, but his books and advocacy simplify a highly nuanced and complex history into five-minute animations like "The Ultimate Roller Coaster Ride." Interpreters might consider borrowing something from the sense of urgency and clarity with which Heinberg conveys the climate crisis and our need to live within our ecological limits.[2] We now know that we cannot stop climate change, but even to adapt, energy users need to help lead the demand to diversify the fossil fuels from which we produce the majority of our energy. Both academic historians and journalists have produced works that can help us understand dependence and choices in context.

Histories and Contexts

The disappearance of woodlands for energy hastened Europeans to enlist fossil fuels for domestic and industrial energy use. Anthracite coal, or "stone coal," soon replaced the wood hearth and the charcoal in the iron furnace. Coal-driven steam provided the energy for early industrial transportation of goods and people. Bituminous and lignite coal, with lower carbon content than anthracite, often generated electricity or operated steam engines.

The stationary steam engine defined working conditions and urban spaces in the industrial era. Developed by small-town British engineer Thomas Newcomen (from a more rudimentary pump designed by Thomas Savery) in the early eighteenth century, it could create mechanical energy to operate machinery. This included pumping to mine subterranean coal. The operation condensed steam into a cylinder. Condensation created a vacuum in the cylinder, resulting in atmospheric pressure that moved a piston. When Scottish mechanical engineer James Watt (for whom we named units of electrical power) and Matthew Boulton further refined the steam engine in 1776 into a more efficient machine, an energy source once desired for single-home heating expanded to the cities. Industrial scientists further perfected technologies that made steam power production increasingly efficient. Such economies of scale fueled the industrial manufacturing that aided America's economic dominance in the late nineteenth and twentieth centuries.[3]

Steam power increased the demand for coal, which accelerated the growth and efficiency of factory work. Machines that could somehow convert the thermal energy from burning coal eventually offered far more energy output per individual worker, and they changed the nature of work itself. With twenty healthy adult males needed to achieve one unit of horsepower, enlisting laborers to operate machines for extraction diminished the need for caloric energy to fuel muscle power. By having people operate cars and manufacturing equipment, "they shifted the burdens of labor off the soil and onto subsoil minerals." This shift essentially fueled the Industrial Revolution, first in Britain, then followed by the rest of Europe and the United States several years and decades later.[4]

Coal had a distinct advantage over other resources. Even when part of a larger system of storage dams and reservoirs, waterpower remained variable with the seasons. The availability of coal, however, only depended on the labor *available* to mine it. Industries dependent on water often had to recruit labor to more remote areas and supply housing and services, but the portability of the steam engine and a fuel source like coal could meet labor supplies where they lived, in the concentrated populations of cities and villages.

Those with economic and social power profited through the muscle energy of the labor classes who extracted coal, natural gas, and oil. Stories about the life of energy workers and their labor is another theme heavily associated with the history of fossil fuels that can humanize energy narratives. The type of labor demanded in the mines and the type of labor that coal-fire steam plants required in factories helped create a multigenerational industrial working class that included men, women, and children. Tapping into the themes of energy workers, and the numerous labor histories and "workscapes" associated with them (see Thomas Andrews's *Killing for Coal*), can further expand interpretative narratives to include the human cost of energy extraction work and the struggle for safe working conditions.[5]

With the aid of technology, energy workers have tapped exhaustive geologic resources acquired from Native land and accelerated Western expansion to build a cheap, but energy-rich, consumer economy and society. With pickaxe and shovel, miners quickly exhausted the anthracite located at the surface of the earth's crust during the late nineteenth and early twentieth centuries. At the same time coal demand grew, coal mining communities became more isolated from urban, industrial places, both physically and in perception. In *Routes of*

Power: Energy and Modern America, Christopher Jones argues that recognizing the spatial distance from sites of production and consumption are critical for understanding our fuel dependence. He stresses that it was the development of transportation networks, including roads, canals, secure oil and gas pipelines, and transmission lines that replaced the work of human transportation (aka teamsters). Fossil fuels are stored in barrels, as well as above-ground tanks or underground in salt or hard rock caverns, mines, aquifers, or depleted reservoirs to ensure steady fuel supplies. This infrastructure allowed most populations cheap and abundant access to energy, building societal dependence on fossil fuels for both domestic and industrial use.[6]

Thankfully some of the most useful and accessible primers for understanding the cultural underpinnings of fossil fuel dependence, and what it might take to transition away from use, include works from the last quarter century written for popular audiences. David Nye's *Consuming Power: A Social History of American Energies* (1998), Ian Morris's *Farmers, Foragers, and Fossil Fuels: How Human Values Evolve* (2015), and environmental attorney Barbara Freese's popular history *Coal: A Human History* (2016) frame the narrative in terms of the relationship between energy systems and social organization and values. In particular, Bob Johnson's *Carbon Nation* (2014) enlists art, literature, and film to examine how deeply fossil fuels have penetrated every aspect of our culture. Each of these works delve into the many ways dependence on carbon-based fuels has altered our relationship with energy, to the point where we overlook the costs of them to our environment and our bodies. We additionally physically and socially hide the process and impact from much of the middle class.[7]

Nye argues the idea that our fossil fuel use was not inevitable, just consistent with and conducive to our political and economic values. He explains that the development of certain energy systems has grown out of value-infused economic systems like capitalism. While capitalism emphasizes competition, the expense of infrastructure limits the number of competing systems that historically led to monopolies like Rockefeller's Standard Oil. Railroads controlled the transportation of coal to market, and likewise oil until pipeline infrastructure challenged that delivery method. Oil began to rival coal as transportation options expanded beyond ships and trains. As automobiles became more affordable and the Federal-Aid Highway Act of 1956 made freight by truck even cheaper and more convenient, transportation via airplanes also increased demand.

The economic beneficiaries of market expansion remained limited to certain segments of the population, often excluding those from whose land oil was extracted. Angie Debo's classic *And Still the Waters Run: The Betrayal of the Five Civilized Tribes* (1940), Kathleen Chamberlain's *Under Sacred Ground: A History of Navajo Oil 1922–1982* (2000), Dana Powell's *Landscapes of Power* (2018), and James Robert Allison's *Sovereignty for Survival: Energy Development and Indian Self-Determination* (2015) explore the tension of oil and gas extraction on Native lands as Native people simultaneously balance holding on to and protecting the integrity of their homeland with finding economic self-sufficiency. See Marjane Ambler's *Breaking the Iron Bonds* for an excellent account of how several tribes formed the Council of Energy Resource Tribes to manage their own energy resources. Against civil rights protests of Red Power and the national story of the 1973 Arab oil embargo, the federal government turned to tribes on resource-rich reservations. The passage of the Indian Self-Determination and Education Assistance Act of 1975 encouraged a revision of the

Indian Mineral Leasing Act of 1938. The earlier legislation had propagated dependence on the federal government for making energy development decisions by requiring competitive bidding procedures and standard lease forms. Ironically, this late twentieth-century political empowerment occurs near the end of the industrial era, but such recognition of Indian sovereignty will continue to be important.[8]

Much of the fossil fuel industry is shrouded, not in government secrecy like nuclear power, but by the proprietary information of private energy companies. Thus, the keepers of this history have predominantly been amateur historians, industry professionals, or journalists, *not* scholars. In the twenty-first century, however, oil and gas industries have received more scholarly attention. Brian C. Black's *Petrolia: The Landscape of America's First Oil Boom* (2003) and *Crude Reality: Petroleum in World History* (2012) and Paul Sabin's *Crude Politics: The California Oil Market, 1900–1940* (2004) helped make headway in scholarly treatment for the industry. Charles Blanchard's *The Extraction State: A History of Natural Gas in America* (2021) is one example that provides a sweeping business- and industrial-focused narrative of that industry.

In 2012, a special issue of the *Journal of American History* edited by Black with Karen R. Merrill and Tyler Priest, aimed to "historicize" oil and its sociocultural as well as economic and geopolitical implications. Journal editor Edward Linenthal realized that as much as oil is in the news, most people, including historians, know very little about it. The special issue's twenty-four essays encompassed geopolitical, visual, religious, and political culture, as well as economic impacts at the local, regional, national, and international levels.[9] Merrill proposed that "Historians' work lies in unveiling and explaining the paths of the past and in understanding the past on its own terms. But with all the evidence surrounding us that climate change is underway, now is the time to begin asking ourselves: what do we most need to know about oil and modern America to explain how we arrived at this new era in earth's history?"[10] Almost every one of these essays identifies reasons that historians and the general public alike have not identified oil as a subject of critical scholarly inquiry. The guest editors noted that the media only covers certain oil-related stories, such as environmentally disastrous oil spills or price jumps. Coal certainly had a more profound impact on how we live and created the initial energy demand and infrastructure, but oil has even more geopolitical importance and is a cornerstone in the growth of capitalism. Like coal (with the exception of those who live in certain regions, towns, and reservations), oil production is largely separated and hidden from those who use and consume it. In the history field, it is primarily business and environmental historians who have paid it any attention.[11]

In 2022, we saw how energy production and exports played a significant part in how the world markets responded to Russia's invasion of Ukraine. Oil production and competition became integral to geopolitics over the last century, when America and much of the world began to depend on foreign oil from South America and the Middle East. This dependence culminated in the 1973 oil embargo and energy crisis that followed. Coinciding with the modern environmental movement, efforts toward a transition to alternative energies like solar characterized the latter part of the decade, but Paul Sabin has argued that the window for change was not open as wide or for as long as many have perceived. While the media's depiction and images around oil, especially after disasters like oil spills, influenced public perceptions, once the crisis passed, environmental concerns tended to fall alongside oil prices. Americans again embraced oil as its preferred energy source, as played out in popular television shows like *Dallas*, which featured a family of oil tycoons.[12] In the current millennium, oil, environmental concerns, and Native sovereignty have come into conflict most glaringly in the cases of the Dakota Access (DAPL) and Keystone XL Pipelines. Native people launched a campaign to argue that the risk to their groundwater supply was both environmental and cultural, framing water as critical to human health. In the case of DAPL, its impact on sacred religious and cultural sites even allowed Section 106 of the National Historic Preservation Act to temporarily halt its construction. The media attention on the protests exposed the stark conflict between protecting the centuries-old environmental and cultural value of Native lands against the economic and cultural power of modern American life.[13] This struggle for control over where and to whom energy goes is not new. It is at the foundation of energy politics and culture.

The history of oil builds upon that of coal's economic importance, but oil also profoundly impacted postwar American life in the explosive growth of the middle class. The accessibility of oil production allowed industry and the government to promote cheap, disposable plastics as consumer products and to make the purchase of those products akin to patriotism. For this reason, Lizabeth Cohen's *A Consumers' Republic: The Politics of Mass Consumption in Postwar America* is a useful reference for arguing that American society chose to adjust to lives of high energy use and dependence was also a choice, but largely not one of individual free will.[14] Ruth Schwartz Cowan describes the competition among gas and oil utilities to capture the home appliance and heating markets and how that played out in home design in *More Work for Mother: The Ironies of Household Technology from the Open Hearth to the Microwave*.[15]

Most historic sites, especially historic houses, keep objects related to lighting within their collections, such as whale oil, camphene (ethanol or ethyl alcohol), kerosene (petroleum), coal (cannel) oil, and gas lamps. Brian Bowers's *Lengthening the Day: A History of Lighting Technology* provides a good guide for understanding that technology. The republished nineteenth-century novel by Edward Watts, *Cannel Coal Oil Days: A Novel*, can provide insight into the rising use of coal oil set against the racial politics and historical events leading up to the Civil War.[16] Linda Simon's *Dark Light: Electricity and Anxiety from the Telegraph to the X-Ray* can provide some insight into how domestic light altered people's sleep patterns, perceptions of time, and relationship to the natural markers of that time. Lighting or heating a home without electrical power was labor-intensive. While kerosene lamps were safer and easier than candles, one had to continually fill them with oil and trim the wicks. Both methods could also cause fires. Indoor light allowed economic and educational advantages by allowing more hours of productivity. Jeremy Zallen's *American Lucifers: The Dark History of Artificial Light 1750–1865* elaborates on the profound impact of lighting on factory production and the nature of industrial work. There are also studies on the impact of light pollution on wildlife ecology, as well as physiological studies on how artificial lights affect sleep deprivation and mental health, cause changes in cultural practices in rural areas, and impact safety in urban areas.[17]

Access to inexpensive (and finite) energy, products, and technology (as well as government programs and subsidies) ironically helped define a modern lifestyle based on a culture of abundance. The arguments in Daniel French's *When They Hid the Fire* and Christopher

Jones's *Routes of Power* that energy production is largely hidden from energy users are persuasive. If we accept the idea that energy use has only increased as we users see and understand less about its production, then making energy production more visible—and accessible—to the public should be the primary focus of our interpretation. When people see energy production and understand how it works, they use less energy.[18] Yet the treatment of fossil fuels at public history sites has focused far more on production and producer, and museums are not connecting those sites to consumers or other resources associated with consumption. Historical interpretation, however, can "strip away the many layers of sedimentation that have accumulated over the years to distance us from the ecological core of our modernity."[19]

Interpreting the Culture of Coal

Several abandoned coal mines have become museums that focus on technology and equipment. However, coal mining also maintains a strong tie to the people who worked in them and the local communities that supported their families and lives.[20] In those places, coal mining developed a treasured heritage that defined all aspects of life: work, family, community, and culture, as intimately tied to extractive economies of energy. Energy extraction pulled such communities along boom-and-bust cycles throughout the twentieth century. The National Coal Association maintains a long list of museums and tours across the United States and Canada.[21] Several of those small, local museums focus on people and the community united through their work. Mining is intimately tied to local stories of immigration, labor, and social history. Author Barbara Freese distinguished coal with the observation that, while we often associate oil with excessive wealth, "coal does not make us think of the rich [like oil], but of the poor."[22] Coal communities are often isolated and insular places, and museums have been slow to connect them to today's energy and labor issues. When editing a journal issue about public history and deindustrialization, historian Christian Wicke wondered if how we present and interpret industrial history might change now that we are experiencing climate change. "Yet, for now," he perceived, "industrial heritage remains closely connected to very local, regional, and national identities."[23]

Having said that, historical treatment of coal history can vary across regions, depending upon how much it is still part of the economy and community or at least conceived as such. Coal-themed museums stretch across Appalachia from western Pennsylvania into Alabama, the upper Midwest, and parts of Colorado and the Dakotas. Many tend to discuss mining technology, without explaining the burning of coal for industrial steam engines. Additionally, a public history initiative of the University of Illinois system observed that:

> These local museums are tremendously important as public history venues because their scripts (materials, organization) are the vision of the people most directly connected with them, rather than museum professionals. Furthermore, they are important because their collections of mining paraphernalia demonstrate the importance that these objects had to the miners who had used them. The collections speak to the strong identity, pride and solidarity of miners as miners.[24]

The Pennsylvania Anthracite Heritage Museum and the Eckley Miners' Village Museum focus some of their interpretation on the technology, but mostly interpret the theme of labor through the lives of the workers and the community as the industry defined it. James Brooks, editor of the journal *The Public Historian*, similarly commented about those in Colorado, "The museums tend to emphasize the diverse ethnic and racial composition of the coal camps, workers' life underground and family life above, and aspects of "welfare capitalism."[25]

Because mining museums are often located in remote industrial communities, they are sites outsiders most often visit as part of a road trip or vacation, and thus "our exposure to these gritty, industrial workplaces usually occurs in the context of recreation or leisure." One of four regional mining museums in the vicinity of Scranton, Pennsylvania, the Eckley Miners' Village Museum has a complete collection of miner houses along an intact streetscape, consistent with the architecture and family life of a company town from 1880. Chosen as the film location for *The Molly Maguires* starring Sean Connery and Richard Harris, the movie studio reconstructed various parts of the landscape. So, while such a place might accurately reflect various characteristics of a historical coal mining town, the architectural and cultural emphasis alone makes the impact of residents' labor on the daily life of energy consumers outside the community, and even the laborers themselves, seem inconsequential, or almost beside the point.[26]

Those who live outside of mining communities may not recognize how mineral extraction impacts the landscapes of those communities because outsiders do not live with the environmental impact. Attempts to preserve and sometimes re-create the mining landscape could open ways to explore the ecological aspects of extracting coal for energy. The muscle energy of miners is necessary for the extraction of coal, and so in this way historic mines and mining towns are also integral to the historical energy narrative. At Eckley, one activity does enlist participants to explore the mineralogy of coal and the process of extracting it. Repositories like the Coal and Coke Heritage Center at Penn State–Fayette (the Eberly campus) provide cultural and economic context for a key source of coal extraction energy—the muscle energy of the laborers to extract, sort, and load coal onto railroad cars to send to urban markets and powerplants. Interpretation could articulate these links between the coal miner and the impact of coal mine work on their body, their family, and the shape and values of their community.[27] Coal communities have felt the economic impact from oil and natural gas competition for the electrical generation market since the early twentieth century.[28] As more alternatives to fossil fuels emerge, museums can become powerful advocates for transition.

The administrative choices and decisions of coal-themed museums can also raise significant awareness about energy transition. The Kentucky Coal Mining Museum made international news in 2017 because of the irony, even to the most casual observer, of the institution switching to solar power for practical reasons, mostly financial. Interpreting that decision process and transition would be the most remarkable contribution this museum could make toward energy education.[29] Likewise, even places not directly associated with fossil fuels can offer broader stories of energy transition.

Interpreting Oil Before and After Petroleum

Places like the New Bedford Whaling Museum, one of several whaling and maritime sites along the coasts, provide context for the vast oil industry that preceded petroleum, much like wood charcoal fuel preceded coal mining. The use and demand of whale oil for lighting from the sixteenth through much of the nineteenth century, the process by which whalers extracted the oil from blubber through heat, and wind energy as the primary sources of energy for whaling ships could all serve as rich interpretive themes. As Heidi Scott explained, "Whale oil culture provides a premonition of this petroleum era paradox of exuberance and catastrophe." Scott noted that "both are made of the bodies of dead organisms," and she described examples from art and literature (such as Herman Melville's *Moby Dick* and Upton Sinclair's *King Coal*) that have celebrated or just romanticized the pursuit of fuels.[30] Maritime museums interpret this cultural history alongside descriptions of natural history and current-day ecologically focused exhibits. Public programming at these sites similarly covers art, science, history, and culture. As suggested previously regarding the Mystic Seaport Museum, the history of whale oil (or plant oils) can serve as a prologue and model for understanding petroleum oil, found in the underground reservoirs of ancient seas.

Maritime museums invite a multidisciplinary approach that integrates technological, economic, and cultural history, but most museums focused on petroleum center around a progress narrative, stressing the critical importance of petroleum to America's energy future and the world's energy security. These interpretive centers remind Americans about the importance of oil, its geologic origin, and how we process it for energy. They celebrate extraction technology, the economic impact, and highlight oil's geopolitical importance. Many have well-developed STEM educational programs. Some are discussing and acknowledging energy diversity, but primarily as supplementary to petroleum. These types of interpretations reflect a deeply seated and widespread cultural assumption that when it comes to reliable energy and growth, there exist no viable alternatives to fossil fuels.

In 2003, about the time climate change activism began to grow more vocal and political, energy reporter and editor Bruce Wells founded the American Oil and Gas Historical Society. In its virtual resource space, Wells explains that the goal of the society is "to educate the public about modern energy challenges."[31] The Petroleum History Institute, also founded in 2003, grew out of an organization affiliated with the Drake Well Museum in Pennsylvania that has hosted symposiums and an annual international peer-reviewed journal. It provides fossil fuel–themed educational materials, and it features a growing database of roadside markers and provides guidance on writing and installing them with official state agencies. According to the websites, nearly half of the states in the union feature museums dedicated to oil and gas exploration, production, and transportation.[32] The regions with the most museums and historic sites devoted to oil and gas development are unsurprisingly the epicenter of development in Pennsylvania, Oklahoma, and Texas. The facilities range from modest to state-of-the-art exhibit spaces, many sponsored and funded by the petroleum and gas industries. While they market themselves as general energy museums, oil and gas technologies and their profound economic (as opposed to environmental) impact serve as the primary themes.

Some places also strongly link interpretation to cultural and historical traditions. The Permian Basin Petroleum Museum in Midland, Texas, contextualizes the story of oil within the frontier pioneer narrative through a facility that moves well beyond a heritage center, reaching multiple audiences in a modern community gathering space. The complex combines the best practices of art, science, natural history, and history exhibit layout, design, technology, and interactivity. Director of the Chamber of Commerce George T. Abell and "500 community leaders" founded the museum in 1975 to "tell the story of petroleum and the rugged lives of men and women who sought a better life."[33] The entire building serves as an integrated site of heritage tourism, with the exterior architecture echoing the motif of a frontier institution, like the presidio, but with a slick, modern interior. Galleries feature geologic exhibits, sophisticated technology and descriptions of processes, and kitsch to emphasize petroleum's profound impact on everyday life. The museum nods toward energy diversity, placing exhibits of solar, wind, hydro, and nuclear power in their own "supplemental energies" exhibit. On the website, the museum urges young visitors to pursue careers in petroleum and energy sciences. Another exhibit geared toward educating young children to conserve energy claims it teaches about "various forms," but the space is sponsored by an oil company.[34] The museum's reach goes far beyond exhibits, boasting several event spaces for the museum and the broader community, a research archives, and educational STEM programs.

Exhibit space also extends beyond the petroleum museum's interior to the forty-acre outdoor "Oil Patch" exhibit of colorful, brightly painted oil rigs and other equipment. The

Figure 5.1. Exhibit, Permian Basin Petroleum Museum, Midland, Texas, 2022. Photo by Jaime Aguila

Figure 5.2.   Outdoor Exhibit, Permian Basin Petroleum Museum, Midland, Texas, 2022. Photo by Jaime Aguila

display lures the visitor out of the pristine space of the museum to the raw, open desert prairie and locates the oil industry upon it. In the *Journal of Museum Education*, museum and park architect Peter D. Paul praised the museum's use of the outdoor environmental landscape as exhibit space and its overall emphasis on "the material culture of place." He observed that, "With the towers of downtown Midland in the background, this display becomes a metaphor for the oil fields and the urban capital they have created." Such a juxtaposition encouraged visitor engagement and inquiry. Paul continued, "These museums present open systems—artifacts but not conclusions, directions but not dictums."[35]

At the same time, the museum site undoubtedly conveys an optimistic message about the future of oil. The application of bright paint colors on these machines set on the dusty prairie against the big sky alters the authenticity of the setting from an oil field into something resembling public art. Inside the museum, one gallery even features similarly framed landscape paintings by artist Tom Lovell. The juxtaposition of those machines on the outdoor landscape also echoes the notion of "the machine in the garden," instinctually encouraging an aesthetic where visitors can easily associate the equipment as a natural part of the landscape, rather than a disruptor of it.

Agricultural settings similarly reflect the coexistence of energy technology on the landscape. Fossil fuels dictate the nature of modern farm work and food production. Petroleum-fueled farm vehicles and equipment freed up land to grow food for market rather than spending money on the muscle energy of agricultural laborers, workers, and draft animals. Food source animals required fewer calories than draft animals. Unfortunately, food production in the United States remains far more energy intensive than in developing countries. In addition to the volume by Michelle Moon in the AASLH series to which the present volume also belongs, she and Cathy Stanton discuss how fossil fuels have generated continuous economic growth and led to overproduction and instability in the food system in their book *Public History and the Food Movement*.[36]

The extent to which fossil fuels have defined our modern life, culture, perspective, and sensibilities, yet remain below the surface of public consciousness, is an interpretive challenge. The work of the Petrocultures Research Group explores this, as does Stephanie LeMenager, author of *Living Oil: Petroleum Culture in the American Century* (discussed in an earlier chapter). LeMenager uses literature and film to archive sensory and emotional values associated with "oil culture." "Petromodernity," and even the idealistic "petrotopia," is characterized by suburbia, highways, and parkways. She also envisions an effective way to have modern visitors confront oil directly when designing a museum around the Los Angeles La Brea Tar Pits, a collection of Pleistocene fossils atop an urban oil reserve (the Salt Lake Oil Field) of a form of petroleum called "asphaltum." LeMenager argues that the museum makes "oil visible as a material presence" that cannot be easily reduced to the abstraction of "energy."[37]

This separation of energy from the environment in public displays of energy has a history. Bob Johnson pointed out that even at places like Philadelphia's 1876 World's Fair, visitors interested in exhibits around energy "preferred to keep their ecological dependencies out of sight" and to indulge instead, as the historian Andrew Goodheart has written, in the more anthropocentric fantasy of technology, progress, and the "perfect-running machine."[38] Without public access to the "boiler room," so to speak, we continue to privilege technology as an interpretive theme. Johnson explains that:

> In order to live with fossil fuels, the nation's middle class learned to bury the human and environmental costs of its dependencies out of sight both geographically and socially in the working-class geographies of the coal mine and stokehole and to repress the many traumas and dislocations associated with fossil fuels both physically and symbolically within a national narrative of progress, emancipation, and empowerment that had little room for modernity's objections and casualties.[39]

Historical displays can reveal to visitors that which has been buried, ignored, overlooked, or forgotten. Exposure can at least help them to start thinking about alternatives and loosening our fossil fuel dependency. Exhibits give visitors the opportunity to more directly bear witness to, and hence be forced to think about, the ecological processes involved in energy production.

CASE STUDY: Drake Well Museum, Titusville, Pennsylvania

From iron furnaces to coal and oil extraction to fracking natural gas from shale rock, Pennsylvania's historical landscape features almost every energy resource, especially in terms of fossil fuel development. Energy development literally drives the place-based heritage tourism in western Pennsylvania. Chris Magoc describes an actual driving tour in an article he wrote for the environmental sustainability edition of *The Public Historian* in 2014. The following is an excerpt from that article, where Magoc discusses how the Drake Museum interprets the birthplace of the American petroleum industry:[40]

> The short drive from timber to oil country is a lovely journey through winding valleys and steeply hilly terrain thickly covered with forest. It was not so in 1865. Shortly after the close of the Civil War, Harper's New Monthly Magazine sent a reporter to the oil region of northwest Pennsylvania already known throughout the industrialized world variously as "Petrolia," "Oildorado," or "Oildom." The reporter described the place thusly:
>
> The soil is black, being saturated with waste petroleum. The engine-houses, pumps, and tanks are black, with the smoke and soot of the coal-fires, which raise the steam to drive the wells. The shanties—for there is scarcely a house in the whole seven miles of oil territory along the creek—are black. The men that work among the barrels, machinery, tanks, and teams are white men blackened. . . . Even the trees, which timidly clung to the sides of the bluffs, wore the universal sooty covering. Their very leaves were black.[41]
>
> As historian Brian Black has written, what we might perceive as a grim characterization of the region was then conveyed first and foremost as a sign of the remarkably explosive growth of the industry that was changing the world. A marvel of the technological-industrial sublime, Oildom's environmentally devastated landscape was celebrated as a dramatic indicator of the extraordinary progress the oil region portended for the nation. Americans embraced what would become a central tenet of their emerging industrial civilization: a willingness to sacrifice local environments for the promise of wealth and an improved material life.[42] The nation received that and more from the "valley that changed the world."[43] In little more than a decade following Colonel Edwin Drake's drilling of the world's first commercially successful oil well in 1859, the Oil Creek valley produced more than fifty-six million barrels of oil, generating $17 million in gross revenue.[44] The region's oil boom led to the birth of the oil industry that transformed the world in ways the colonel could have never imagined.
>
> Today that story is told at the Drake Well Museum in Titusville, a Pennsylvania Historical and Museum Commission-administered site (and the more modest Venango Museum of Art, Science and Industry in Oil City). Drake Well's permanent exhibition, opened in 2012, bears the title There's a Drop of Oil and Gas in Your Life Every Day! After taking visitors through the geologic origins and constitution of petroleum, the exhibition explores Native American uses of oil, early innovative "Petroleum Pioneers,"

and the evolution of drilling techniques. The exhibition is multi-layered and engaging, examining the interplay of technological development, the growth and social disruption of the regional boom-bust oil economy, irresistibly compelling incendiary disasters that engulfed entire communities in fire, the growth of the industry well beyond Pennsylvania, and the economic manipulation of the industry led by John D. Rockefeller. Artifact-rich, the installation also features dramatic images, extensive primary source material, compelling dioramas, and an overall effective interactivity.

Further, Drake Well delivers masterfully on its central theme: the transformational power and dominance of oil in modern life. As one panel declares: "Oil and natural gas have forever changed how we work, where we live, how we travel—maybe even the air we breathe. What would life be like without them?" Visitors learn of the industry's virtually continuous process of scientifically engineered innovation that has delivered much of our material world. "What can we make from oil and gas?" one panel asks. "Today, thanks to the petroleum chemistry, our lives include synthetic medicine, manufactured wood, and sticky notes."[45] Nostalgic images of lustrous 1950s cars and road-side architecture celebrate the American love affair with cars that is at the heart of oil's story. The possibility of life without oil, we conclude, is simply unthinkable.

The exhibition's final sections address some of the hard realities of the oil age. They do so in part by necessity: given the limitations of gallery space, they collapse the last four or five decades of geopolitically complex and increasingly violent oil history with summary statements. For example, we learn that "because petroleum is the 'blood' of military power, dependence on foreign oil can become a national security issue. When supplies are threatened, countries go to war to protect their sources." Although both Drake Well and the Venango Museum arguably do well simply to acknowledge the connection between fossil fuels, the American way of life, and at least vaguely, global U.S. military power, this sort of opaque phrasing is not especially enlightening or useful in thinking about today's challenges. PHMC staff intends to more fully develop this immense story in an expanded future gallery. This is encouraging, for the need to offer American museumgoers a candid exploration of the manifold historic and contemporary connections between oil, war, and the growth of the U.S. military establishment seems urgently needed and long overdue.

A concluding gallery at Drake Well braves the cold geophysical facts of diminishing fossil fuel supply and illustrates the ongoing search for a more diversified energy future. Here we also see evidenced a primary educational objective of PHMC staff: to have visitors "look at their consumption,"[46] and to convey to visitors their own role in the statistics of depletion. As one panel puts it bluntly, "the U.S. produces 7 million barrels a day; we use 20 million barrels." The exhibit urges "Recycle, Reduce, Reuse," and then offers a brief survey of alternative technologies that can supplement, though one deduces never replace, the central role of oil in our society. With visitors imagining "What will our energy future be?" their answer gropes for something more satisfying than the exhibition can honestly offer: "Although the search for oil and natural gas may have caused acid rain, oil spills and global warming, it is hard to imagine a completely petroleum-free future." Indeed it is, particularly after the exhibition's tour of an alchemical wonder world of utopian plastics, convenience, leisure, and mobility from which no rational person would ever want to escape. The museum settles for what feels like an unsettling mix of sanguine reassurance and ineffably terrifying implications.

The Venango Museum of Art, Science and Industry reinforces that ambivalent theme through a short film. Funded by the American Petroleum Institute, FUEL-LESS follows "Crystal," a young teenager who wakes up one day to discover, much to her horror, that she is "FUEL-LESS": no hair dryer, no makeup, no gas, and her dress is a burlap sack. Crystal's teacher enlightens her as to the means by which oil is refined and magically transformed into the phenomenal products of her existence. The film urges individual responsibility and assures that commensurate corporate stewardship is well in hand: Crystal is told that "new standards have been adopted" to prevent the icky "oil spills [she] used to see on television," and that bacteria can now eat any oil that somehow might—you never know—get into the ocean.[47]

The oil-blackened landscape of 1865 Oildom is not in evidence on the pristine and tree-covered landscape of Drake Well grounds today. Here one sees blacksmiths hammering at replicated small rigging operations—a restored verdant heritage landscape that belies even the despoiled place of yesteryear pictured in the museum. Such quaint verisimilitude of the glory years is delightful and educational, but I could not help thinking how it serves to further insulate visitors from present-day threats associated with the same industry: the horrors of oil refineries in places like Nigeria, for example, or the climate crisis to which our dependence on sticky notes continues to contribute.[48]

Every time I drive away from Oildom in my own gas-powered machine, I find myself asking this question: how much can one reasonably ask of a museum situated in the heart of historic oil country? Drake Well Site Administrator Melissa Mann and Curator-Historian Susan Beates argue that the museum needs to "meet people where they are," which, in their charitable characterization, is "uninformed" about even the source of their energy—never mind the global climate-changing, defense budget-swelling implications of their consumption. Mann and Beates hope the site not only educates them about the history and sources of American energy supply, but also "strengthens their ability to think critically" about what they hear on television and read on the internet about energy-related issues.[49] That mission is extended through a traveling educational program called "Mobile Energy Educational Training Unit" (MEET-U). A forty-four-foot-long trailer of exhibits that has traveled to schools and energy industry events throughout a three-state region, MEET-U educates children about the history, technology, and contemporary challenges of America's energy supply. Approximately seventy-five percent of the funding for this program comes from the fossil fuel industry. Mann and Beates state without hesitation that the connection between the industry and Drake Well ends there. Moreover, they are adamant that Drake Well does "not serve as the mouthpiece for the industry."[50] They cite the fact that source material for MEET-U and portions of the exhibition at Drake Well derive from the U.S. Department of Energy (DOE)—which they assert is a scientifically unimpeachable, objective, and independent source of information on all issues energy-related. That assumption appears to discount the historically entwined relationship of the DOE to the industry. The appointment in 2013 of Ernest Moniz, a physicist with deep connections to the oil and gas industries, as DOE Secretary reflects what agency critics have long described as an industry-cozy relationship.

How much can one reasonably ask of a local museum in an area with a complex environmental and industrial history? What is the purpose and function of a local or regional museum when it comes to important questions of natural resources and the environment? These questions, at the heart of this essay, were left uneasily unanswered for me here. PHMC staff, recalling that Drake Well was established in 1934 by the American Petroleum Institute (API), proudly assert the evolution of the site from what was largely an API organ that unabashedly promoted the interests of the oil industry to a more scientifically well-grounded educational institution whose mission is to educate and challenge its visitors to think critically. Compared to what I recall from a visit to Drake Well in the early 1990s, the museum has indeed come a long way in its interpretation. However, my conversation with staff suggested that for those of us asking for more,—perhaps too much—Drake Well's mission to stoke critical thinking remains somewhat hamstrung and circumscribed by both the institution's location in the historic heart of the industry's birth, and by the deeper politics of contemporary energy extraction in the United States. Beates offered sweeping condemnation of Josh Fox's controversial Gasland film. Moreover, she asserted that fracking opponents are simply and grossly misinformed about the geology and technological prowess of the practice that allows the industry to safely extract this essential fuel from deep in the earth. Mann acknowledged that "there are problems with the industry," but it is important, she said, to "not judge the industry on what we do not know."[51] History—and, we might hope, some future Drake Well exhibit—will render conclusions on the ramifications of an industry that has drilled more than 80,000 hydro-fracturing wells across the country since 2005 and produces more than 200 billion gallons of toxic wastewater every year.[52]

Moreover, it seems important to acknowledge that at least part of "where people are" right now on the issue of fracking is worried and angry: the Pennsylvania Department of Environmental Protection received more than two thousand complaints from Pennsylvania residents related to the industry between 2011 and early 2014, most of them having to do with alleged water contamination.[53] It seemed impossible to not conclude that the site, despite an impressive installation and genuinely asserted confidence in the objectivity of its interpretation and educational programming, remains at least partially if unconsciously captive of the long reach of the interests that birthed Drake Well some eighty years ago.

Interpreting Natural Gas

When the Forest County Visitor Center in western Pennsylvania enlisted Magoc to help develop interpretive signage, he noted that while the narrative highlighted themes of environmental stewardship following deforestation, it avoided the more modern-day environmental controversies around oil and gas extraction. The signage included only statements that gas in the Marcellus Shale region was driving an industrial resurgence and did not address the questions surrounding the energy source and the process of fracking.[54] Gas development is a less culturally pervasive phenomenon than oil, but not new, although the topic has not received broad historical interpretive attention. One exception is the Stevens

County Gas and Historical Museum in southwestern Kansas, featuring one of the world's largest natural gas fields and the history of its development in the 1920s and 1930s. Like the multisite Oil and Gas Museum in Parkersburg, West Virginia, the Kansas site's interpretation is fairly traditional, is heavily artifact-based, and emphasizes the technology. It proudly and legitimately celebrates the developers as persevering community founders and praises the local economic impact of their enterprise.

There are other opportunities for examining regional gas development at historic sites. In a 2012 blog entry called "Don't Frack Our History," then–graduate student Jeff Robinson suggested that public historians should address the latest controversies in hydraulic fracturing (aka fracking), a method of extracting natural gas by introducing pressurized fluids into rock layers. The northern tier of Pennsylvania sits upon the Marcellus Shale, a rock formation that contains the rich natural gas reserves that many claim are needed for America's energy future. When Cabot Oil and Gas tried to acquire land leases at historic places in the early twenty-first century, including some along the Underground Railroad, young protesters marched and cited a different past when big industry left many families "in hardscrabble poverty." Robinson pointedly asked, "How do we bring both the diversity of opinion and the question of specifically politicized values into our public history work, especially at sites and discourses where energy development, climate change, corporate exploitation, and agricultural shifts are prevalent?" He advocated that the controversy could open a dialogue outside of political interests. Scientists and engineers know little about the history and relationship of people to the landscape, and historians do not always acknowledge the scientific intricacies. "If space for serious, mutual relationships and conversations can exist between people with differing opinions and values, then perhaps compromises can prevail," he commented, envisioning "public historians acting as *citizen historians* in forging those dialogues."[55]

One way to provoke such dialogue around energy is to critique and contextualize the interpretation of even very traditional exhibits such as dioramas. In a piece for an energy-themed journal, Samantha Boardman described what one modest tourist attraction can convey about energy, through a limited window in time:

Along the rolling, bucolic stretch of I-78 between Allentown and Harrisburg, billboards entice travelers to exit at Shartlesville for "Roadside America: The World's Greatest Indoor Miniature Village." A local institution since 1953, this attraction features remarkably detailed, hand-crafted, miniature scenes of American history, industry, and progress, arranged in a sweeping, eight-thousand-square-foot tabletop tableau.[56] The life's work of creator Laurence T. Gieringer, Roadside America, with its emphasis on models of regional landmarks and locales, serves as a multifaceted material-culture "text" through which to explore key relationships between energy sources and Pennsylvania's lived history.

While scenes of early frontier settlements show the use of water-, horse-, and manpower, it is in the richly detailed depictions of the coal, petroleum, and electricity industries that Roadside America shines—literally.

Particularly striking is the scale model, "sponsored" by the Reading Iron Works, of the Philadelphia and Reading Anthracite Colliery.[57] This marvelous miniature features a cross section of the mine's tunnels as well as a replica of the Locust Point coal breaker, which was, at the time of the model's construction, the largest of its kind in the world. The rail yard abutting the coal works emphasizes the interconnectedness of the coal industry in Pennsylvania with the country at large, as tiny cars wait to be filled with the extracted anthracite and race across tracks spanning the length and breadth of the model. Likewise, the oil refinery model harks back to the dawn of the American petroleum industry at the Drake Well in Titusville. The miniature Esso filling station with automobiles lining up at its pumps, located in the downtown section of the village of "Fairfield," illustrates the connection between the fuel and its uses.

In Roadside America, electricity is presented as a marker of modernity and progress. The miniature power plant, touted in the tour brochure as having "every brick . . . handcarved in complete detail," is situated at the center of the display, from which it appears to provide the energy to power the brightly lit movie theatre marquee in Fairfield as well as the interior lights of the village's residential and commercial districts.[58] This illumination is brilliantly displayed during the "Night Pageant" that occurs every twenty minutes, in which the lights in the room housing the model are dimmed in a simulated sunset. As Kate Smith's "God Bless America" plays over the public address system and pictures of Jesus, angels, and patriotic scenes project onto a back wall near a fluttering American flag, the sections of the model representing modern America blaze brightly while colonial and pioneer scenes fall into darkness.

The dynamic models of Laurence T. Gieringer's Roadside America vividly depict energy development and its numerous manifestations throughout central Pennsylvania's history. The interconnectedness of energy and history is perhaps nowhere better illustrated than in the model trains that crisscross the model. Steam, electric, and diesel engines share the same tracks, racing through towns whose streets teem alternately with horse-drawn wagons and automobiles, telegraph wires and telephone poles. These anachronistic juxtapositions underscore the technological development enabled by harnessing these energy sources while connecting such innovations to the land and people from which they derived. This optimistic depiction of energy development and applications is due in part to the particular moment in history captured by the landscape of the attraction. After Gieringer's death in 1963, no additional models were added to the display, making the attraction a virtual time capsule of midcentury America. Thus, an unambiguous narrative of progress is not complicated by more recent developments, such as the disasters at Centralia and Three Mile Island or controversy surrounding the fracking of the Marcellus Shale. Likewise, the unintended consequences of transportation fuel innovations are, literally, outside the borders of the display; the America of Roadside America is absent suburban sprawl, deforestation, or mountaintop removal mining. Gieringer's America is one of innovation and potential—a tableau of a promise that had yet to be broken.[59]

Most transportation museums, however, feature life-size artifacts. They are often limited in the size of their artifacts, constrained by donors enthusiastic about technologies, as well as visitor expectations. Because of their space requirements, they are often located on the outskirts of large population centers and attract a niche audience. The size of transportation museum spaces offers opportunity for adjacent narratives through film screenings, public events, and programming that might provide additional context about energy choices, but also target additional demographic groups.

Artifact Spotlight

Automobiles

While almost half of the states in the union feature petroleum and gas museums, almost every state has some type of transportation museum. These spaces are usually packed with automobiles from multiple eras. Such artifacts can encourage visitors to consider political and economic, as well as technological, reasons for why the internal combustion engine gained favor, and thus why petroleum gasoline became the preferred automobile fuel over alcohol (a biofuel), steam, biofuel, or electricity. Cars have a large, nostalgic, and committed hobbyist following interested in design and technology, and enthusiastic volunteers often fund and operate museums holding those collections. Even the Permian Basin Petroleum Museum looked to broaden visitor attendance by expanding its collection beyond oil drilling and refining to include vehicle displays. However, such places can engage an even wider audience with broader interpretation. Since motor vehicles are the most notable consumers of oil and producers of carbon emissions, such exhibit spaces can approach their discussion of engines through the lens of energy.

When the theme of energy is submerged in favor of technology, alternative fuel engines become mere curiosities. Exhibit text writers need to take care not to portray alternative fuel vehicles as only quirky footnotes to a primary narrative. The Owls Head Transportation Museum in Maine played with and reflected the promise and peril of this kind of presentation in its *Fads and Failures* exhibit. Diminished resources scaled the exhibit down from initial plans that focused more on energy competition, but the theme still offers an intriguing framework for examining innovation while engaging visitors in critical thinking about marketing and consumer choice. Here, the introductory panel placed "progress" in quotations, implying (and at the same time questioning) that the alternative fuel innovations of Francis Woods's 1916 Mobilette Roadster, the 1903 Prescott Steam Runabout, or the miniature steam engines of the De Dion-Bouton Voiture company were not a part of the *actual* progress, defined as that which succeeded in the marketplace. Yet without exploring the reasons and context for the transportation industry's embrace of the gasoline-powered combustion engine, the historical fate of alternative fuel vehicles continues to be sidelined in the minds of today's consumers. A more explicit exhibit theme of energy could likewise provide a

Figure 5.3.   Image Owl Head Transportation Museum "Fads and Failures" exhibit, Owls Head, Maine, 2021. Photo by author

context for even vehicle enthusiasts to consider alternative fuel vehicles as not just something for the future, but a return to past ideas and innovations.[60]

To broaden the interest of potential visitors beyond hobbyists, several transportation museums are reinterpreting collections with perspectives beyond technological history that expand audience interest, such as through the social context of race or gender. As exhibits change at the Seal Cove Auto Museum, located on Mount Desert Island in Maine, the artifact vehicles rotate around the space. Their interpretation adapts to a changing exhibit theme, such as 2020's focus: the role of cars in women's suffrage.[61]

Petroliana

This category of material culture generally refers to gas station equipment and advertising kitsch from oil companies, but collectibles include gas pumps, oil cans, road maps, and signs from Mobil, Texaco, Standard Oil, Phillips 66, Shell, Sinclair, and Esso.[62] The display and interpretation of these items, which are generally easy to acquire at yard sales or flea markets, can encourage visitors to think about the relationship between petroleum and consumerism. This connection is becoming more critical as consumers prefer internet shopping and free delivery, perpetuating the role of fossil fuels in the low-cost transportation of goods.

Figure 5.4. Exhibit featuring "petroliana," Permian Basin Petroleum Museum, Midland, Texas, 2022. Photo by Jaime Aguila

Household Lamps

Lighting fuel was as much of a cultural, economic, and political choice as it was a technological decision. Lighting oils tell a micro-story about the shift from biofuel to fossil fuel. Interpretation can address such practicalities of use, as well as where people acquired their lighting fuels, the quality of light each fuel type could emit, and how and why the home's inhabitants needed and used that lamp. For example, kerosene lamps were often cheaper than gas, did not need access to gas service, and generally provided steadier illumination. One could also carry them from room to room. After the Civil War, the petroleum industry eventually dominated indoor lighting until the incandescent light bulb (often powered from a distant coal-fired power plant) hit the markets in the 1880s.

Many historic sites and museums can enlist the electric wiring systems in their own buildings or visitor centers as artifacts for energy interpretation. Buildings constructed or renovated after 1930 physically, structurally, and spatially accommodate the fossil fuel infrastructure (furnaces, stoves, radiators) needed for heating and cooking. Household layouts and technologies, especially in homes built at the turn of the century, reflect the debates about fuel and the evolution from hearths to central heating systems powered by coal, oil, gas, or steam from a distant power plant. (See the chapter on wood for the role of stoves in interpreting changes in home heating and cooking, and the chapter on electricity and infrastructure for more about how central oil and gas furnaces and electrical systems impacted house design and chores.)

For sure, public programming around exhibits can help amplify themes of energy choice and diversity, but fossil fuels seeped deep into every crack of the economy, politics, and culture of developed nations. Coal, oil, and gas have fueled economic growth, along with economic inequality. As a later chapter will explain, the infrastructure we have built to support the transportation and delivery of these fuels to industries and consumer markets has dictated the organization and socioeconomic hierarchy of our communities in the twentieth century. Fossil fuels are so prevalent that transitioning to cleaner, renewable energy sources will arguably be more of an economic, cultural, and emotional hardship than a technological or intellectual one.

Notes

1. In 2014, NCPH interviewed Heinberg in its Public Plenary to explore the theme of a post-carbon future for its theme "Sustainable Public History." *Public History News* 34:2 (March 2016), 2.
2. Richard Heinberg, *Blackout: Coal, Climate, and the Last Energy Crisis* (Gabriola Island, BC Canada: New Society Publishers, 2009); Tyler Priest, "Sound Bite History Reconsidered," January 27, 2014, https://ncph.org/history-at-work/sound-bite-history-reconsidered.
3. Nye, *Consuming Power*, 122–23, 131–54; See also Jones, 123–60.
4. Bob Johnson, *Carbon Nation: Fossil Fuels in the Making of American Culture* (Lawrence: University of Kansas Press, 2014), 7, 41.
5. Andreas Malm, *Fossil Capital: The Rise of Steam Power and the Roots of Global Warming* (Brooklyn, NY: Verso, 2016), ebook, chapter 7; Barbara Freese, *Coal: A Human History* (Cambridge, MA: Perseus Publishing, 2003), 78–79. Also see Thomas G. Andrews, *Killing for Coal: America's Deadliest Labor War* (Cambridge, MA: Harvard University Press, 2008).
6. Christopher Jones, *Routes of Power: Energy and Modern America* (Cambridge, MA: Harvard University Press, 2014).
7. Also see Johnson, *Carbon Nation*; Ian Morris, *Foragers, Farmers, and Fossil Fuels: How Human Values Evolve* (Princeton, NJ: Princeton University Press, 2015); and Ross Barrett and Daniel Worden, eds., *Oil Culture* (Minneapolis: University of Minnesota Press, 2014).
8. Angie Debo, *And Still the Waters Run: The Betrayal of the Five Civilized Tribes* (Princeton, NJ: Princeton University Press, 1973); Kathleen Chamberlain's *Under Sacred Ground: A History of Navajo Oil 1922–1982* (Albuquerque: University of New Mexico Press, 2000); James Robert Allison, *Sovereignty for Survival: Energy Development and Indian Self-Determination* (New Haven, CT: Yale University Press, 2015).
9. Darren Dochuk, "Blessed by Oil, Cursed with Crude: God and Black Gold in the American Southwest," *Journal of American History*, 99:1, June 2012, 51–61, https://doi.org/10.1093/jahist/jas100; Myrna Santiago, "Culture Clash: Foreign Oil and Indigenous People in Northern Veracruz, Mexico, 1900–1921," *Journal of American History*, 99:1, June 2012, 62–71, https://doi.org/10.1093/jahist/jas114.

10. Karen R. Merrill, "The Risks of Dead Reckoning: A Postscript on Oil, Climate Change, and Political Time," *Journal of American History*, 99:1, June 2012, 252–55, https://doi.org/10.1093/jahist/jas101.

11. Ed Linenthal with Brian C. Black, Karen R. Merrill, and Tyler Priest, "Oil in American History," JAH Podcast (Journal of American History, Organization of American Historians, June 2012), accessed https://jah.oah.org/podcast, December 1, 2021. In his book *Finding Oil: The Nature of Petroleum Geology 1859–1920* (Lincoln: University of Nebraska Press, 2011), historian Brian Frehner covers both business and environmental history. He explores early efforts in the late nineteenth and early twentieth century to find oil and the evolution of petroleum geology as an intellectual and scientific field, but also an industry of "oil men" who sought power.

12. Kathryn Morse, "There Will Be Birds: Images of Oil Disasters in the Nineteenth and Twentieth Centuries," *Journal of American History*, Volume 99, Issue 1, June 2012, 124–34, https://doi.org/10.1093/jahist/jar651; Karen R. Merrill, "Texas Metropole: Oil, the American West, and U.S. Power in the Postwar Years," *Journal of American History* 99: 1, June 2012, 197–207, https://doi.org/10.1093/jahist/jas096; Paul Sabin, "Crisis and Continuity in U.S. Oil Politics, 1965–1980," *Journal of American History* 99:1, June 2012, 177–86, https://doi.org/10.1093/jahist/jas086.

13. American Cultural Resources Association, "Statement on the Dakota Access Pipeline Controversy," September 28, 2016, accessed January 7, 2022. https://www.acra-crm.org/resources/Pictures/ACRADAPLStatement_9_28_2016.pdf.

14. Brian C. Black, "Oil for Living: Petroleum and American Conspicuous Consumption," *Journal of American History* 99: 1, June 2012, 40–50, https://doi.org/10.1093/jahist/jas022.

15. Cowan, *More Work for Mother*, 89–91.

16. Theophile Maher with Edward Watts (ed.), *Cannel Coal Oil Days: A History* (Morgantown: University of West Virginia Press, 2021).

17. Brian Bowers. *Lengthening the Day: A History of Lighting Technology* (Oxford University Press, 1998); Paul Sabin et al. "Energy History" (Yale University), https://energyhistory.yale.edu/units/harvesting-light-new-england-whaling-nineteenth-century.

18. Daniel French, *When They Hid the Fire: A History of Electricity and Invisible Energy in America*; Jones, *Routes of Power*; "In Plain Sight: Stories of American Infrastructure," *Backstory* (June 5, 2015), accessed January 10, 2022, https://www.backstoryradio.org/shows/infrastructure.

19. Johnson, *Carbon Nation*, 3.

20. See Richard Francaviglia, *Hard Places: Reading the Landscape of America's Historic Mining Districts* (Iowa City: University of Iowa Press, 1991).

21. National Mining Association, "Mining Museums and Tours," https://nma.org/about-nma-2/resources/mining-museums-and-tours, accessed September 27, 2021.

22. Freese, 2.

23. Christian Wicke, "Deindustrialization in Historical Culture," *History@Work* (November 15, 2017), https://ncph.org/history-at-work/deindustrialization-in-historical-culture.

24. University of Illinois System, "Coal Mining Museums: Mythic Mississippi Project: Cultural Heritage Promoting Community Development," accessed December 20, 2021, https://mythicmississippi.illinois.edu/coal/coal-mining-museums.

25. James F. Brooks, "Rust, Recreation, and Reflection," October 27, 2017, https://ncph.org/history-at-work/rust-recreation-and-reflection; also Introduction "Deindustrialization," *The Public Historian* 39:4 (November 2017).

26. Francaviglia, 69–70 100–01, 190–91, 205.

27. Johnson, *Carbon Nation*, 62–69.

28. David Nye, *Consuming Power: A Social History of American Energies* (Cambridge, MA: MIT Press, 1997), 196–97.

29. "The Coal Museum Switched to Solar," Morning Edition National Public Radio, April 7, 2017. Sarah Sutton addresses many other ways historical institutions can interpret their own sustainable decisions regarding operations as part of the interpretation of their site in *Environmental Sustainability at Museums and Historic Sites* (Lanham, MD: Rowman and Littlefield, 2015), 117–48.

30. Heidi Scott, "Whale Oil Culture, Consumerism, and Modern Conservation," in Barrett and Worden, *Oil Culture*, 5–7.

31. David Wells, American Oil and Gas Historical Society, https://aoghs.org, accessed July 21, 2021.

32. Petroleum History Institute/Sam T. Pees, "Oil History," http://petroleumhistory.org/OilHistory/pages/org/PHI.html, accessed September 25, 2021.

33. The Petroleum Museum, https://petroleummuseum.org/contact/about-the-museum. While the pioneer theme is most common in the West, the museum in western New York is actually called the Pioneer Oil Museum.

34. The Petroleum Museum, https://petroleummuseum.org/about-the-museum, accessed July 19, 2021.

35. Peter D. Paul, "The Civic Museum: A Place in the World," *Journal of Museum Education* 24:1 (1999), 6–10.

36. Johnson, 36–39; Michelle Moon, *Interpreting Food at Museums and Historic Sites*; Michelle Moon and Cathy Stanton, *Public History and the Food Movement: Adding the Missing Ingredient* (New York: Routledge, 2018), 67–69, 70–72, 77–78.

37. Stephanie Lemenager, "Fossil, Fuel: Manifesto for the Post-Oil Museum," in *Oil Culture*, 2014, 311; and *Journal of American Studies* 46: 2 (May 2012), 375–94; LeMenager, "The Aesthetics of Petroleum, after Oil!" *American Literary History* 24: 1 (Spring 2012), 59–86; Adam Carlson, Imre Szeman, and Sheena Wilson, *Petrocultures: Oil. Politics, Culture* (Montreal: McGill-Queen's University Press, 2017). See also Imre Szeman and Jeff Diamanti, *Energy Culture* (Morgantown: West Virginia University Press, November 2019).

38. Johnson, xviii, as quoted in Andrew Goodheart, "The Myth and the Machine," *Design Quarterly* 155 (Spring 1992): 24–28.

39. Johnson, xviii, xx.

40. From Chris Magoc, "Reflections on the Public Interpretation of Regional Environmental History in Pennsylvania," *Public Historian* 36:3 (August 2014), 59–65.

41. *Harper's New Monthly Magazine* 50 (1865): 54; in Brian Black, *Petrolia: The Landscape of America's First Oil Boom* (Baltimore: Johns Hopkins University Press, 2000), 66–67.

42. *New York Times*, December 20, 1864; quoted in Black, *Petrolia*, 71; and Black, 79. This kind of cheery celebration of a degraded landscape is fairly typical of the age. See, for example, Chris J. Magoc, *Yellowstone and the Creation of an American Landscape* (Albuquerque: University of New Mexico Press, 1999), chapter 3.

43. The title of a documentary produced by WQED in Pittsburgh and also the tourism moniker claimed by Titusville, Oil City, and the broader "Oil Heritage" region.

44. Black, *Petrolia*, 5.

45. All Drake Well Museum quotations are from the permanent exhibition *There's a Drop of Oil and Gas in Your Life Every Day!* (Harrisburg: Pennsylvania Historical and Museum Commission, Bureau of Historic Sites and Museums and Hilferty and Associates, 2012).

46. Melissa Mann, Drake Well, site administrator, and Susan Bates, curator and interpretive specialist, interview with Chris Magoc, April 17, 2014, in Magoc, "Reflections," 61–62.

47. 'Here Today, Gone Tomorrow?" at Venango Museum of Art, Science and Industry (Houston, TX: Kingsley Productions, 2001).

48. Hari M. Osofsky, "Climate Change and Environmental Justice: Reflections on Litigation Over Oil Extraction and Rights Violations in Nigeria," *Journal of Human Rights and the Environment* 1, no. 2 (September 2010): 189–210.

49. Mann and Bates, April 17, 2014, interview with Magoc.

50. Mann and Bates, April 17, 2014, interview with Magoc.

51. Mann and Bates, April 17, 2014, interview with Magoc.

52. Alison Sider, Russell Gold, and Ben Lefebvre, "Drillers Begin Reusing Frackwater," *Wall Street Journal*, November 20, 2012, http://online.wsj.com/news/articles/SB100014240 52970 20393700457807718311240960, accessed April 19, 2014.

53. Naveena Sadasivam, "In Fracking Fight, a Worry About How Best to Measure Health Threats," *ProPublica*, April 1, 2014, http://www.propublica.org/article/in-fracking-fight -a-worry-about-how-best-to-measure-health-threats, accessed April 19, 2014. For a recent scientific summary of what the authors describe as "major uncertainties" regarding possible water contamination and other public health threats posed by the hydrofracturing of natural gas and oil, see John L. Adgate, Bernard D. Goldstein, and Lisa M. McKenzie, "Potential Public Health Hazards, Exposures, and Health Effects from Unconventional Natural Gas Development," *Environmental Science and Technology*, February 24, 2014, http://pubs.acs.org/ doi/abs/10.1021/es404621d, accessed April 19, 2014.

54. Magoc, 58.

55. Jeff Robinson, "Don't Frack Our History," *History@Work*, September 28, 2012, https://ncph .org/history-at-work/dont-frack-our-history-using-the-past-for-environmental-activism-in -northeastern-pennsylvania, accessed September 27, 2021.

56. Peter George, "Roadside America: An Institution along Route 78," *Village Chronicle*, n.d., 56, clipping courtesy of Dolores Heinsohn personal archive. Though the current location of Roadside America in Shartlesville dates to 1953, the model has long been a regional sensation, being publicly exhibited in one form or another since 1935.

57. Agius, *Story of Laurence T. Gieringer*, 77.

58. Roadside America, Inc., *Pennsylvania's Roadside America Incorporated: The World's Greatest Indoor Miniature Village*, n.d.

59. This essay was originally printed in Samantha Boardman, "Roadside America and the Engine(s) of Progress," *Pennsylvania Magazine of History and Biography* 139:3 (October 2015), 363–65.

60. The OHTM also features an "energy" room, anchored by an 1885 Corless steam engine that serves as its primary space for education. While the current exhibit follows a traditional "from horses to horsepower" technological history, the curator hopes to update it. Conversation (via Zoom) with Rob Verbsky, Curator, OHTM, June 28, 2022.

61. "Engines of Change," Seal Cove Auto Museum, Seal Cove, Maine, 2020–2021, visited July 13, 2021.

62. "Petroliana," accessed December 1, 2021, https://www.collectorsweekly.com/petroliana/ overview.

CHAPTER 6

National Security and Alternative Energy

Nuclear Power During and Beyond the Atomic Age

PUBLIC VIEWS OF NUCLEAR ENERGY differ from other energy sources. In a *New York Times* article about his visit to Chernobyl in Ukraine (the site of the world's largest nuclear disaster in 1986), Mark O'Connell admitted, "I was on a kind of perverse pilgrimage: I wanted to see what the end of the world looked like."[1] Due to its association with the atomic bomb and with dangerous and tragic accidents like at Chernobyl, historical interpretation around nuclear energy has primarily lived at science museums or historic sites associated with danger, tragedy, and violence. Thus, in addition to celebrating technological innovation, many nuclear energy–related sites have become part of "Dark Tourism," a term J. John Lennon and Malcom Foley coined to describe the act of visiting sites that primarily recall death and suffering. These include concentration camps and battlefields, as well as sites of nuclear testing, production, weapons, contamination, meltdown, waste disposal, and tailings.[2] Morbid public interest in nuclear tourism can, however, undermine the historical context, inquiry, and understanding of nuclear energy as an interpretive goal, not to mention public health and safety. In her article looking at "dark public history," Jessica Moody suggested that while public historians need to connect to visitor interests, we must balance the grim and morbid fascination with interpretation that should "contextualize, complicate, and politicize."[3]

Oddity and mystery may also contribute to the appeal of nuclear energy sites, both current and historic. Nuclear energy development, and its proliferation for military weapons and industrial power use, remains unknown to many. The history and development of nuclear, aka atomic, energy began as a top-secret and highly confidential exercise of the military-industrial complex. Sites like the Nevada Test Site "became, in a very literal sense, a buried archive, sedimented layers of technoscientific history waiting to be revealed."[4] Similarly, sites of nuclear power plants require interpretation that aids public understanding. While not numerous, these sites are important to discuss as part of helping the public understand energy history. Furthermore, nuclear energy, in its testing and potential effects on the health and well-being of human and environmental surroundings, remains an important historical topic to interpret beyond these specific sites.

A properly operating reactor uses the energy released during nuclear fission to heat water into steam. The steam provides pressure to spin turbines that generate electrical current. The reaction rate must be continuously monitored and controlled by technically complex systems. A catastrophic failure of the cooling and safety systems could lead to an overheating ("meltdown") of the fuel and release large amounts of radioactive elements, which has invited understandable concern from scientists and the public alike.[5]

With the decommissioning of nuclear power plants, interpreting and preserving nuclear cultural heritage is a growing field, particularly in Europe where plants are aging. A consortium of universities and museums in the United Kingdom, Sweden, and Lithuania has developed a project known as NuSPACES to study this field. NuSPACES shares ideas and best practices for cultural activities around nuclear research, interpretation, preservation, and developing sustainable governance for these sites.[6]

This exercise must address all kinds of cultural complexities, but also technological and political ones. Domestic nuclear power programs have roots in a powerful technology originally developed for a military purpose. For some nations, that is still the case. The United States and its allies have long suspected places like Iran and North Korea of building up a nuclear weapons arsenal under the guise of a domestic power program. Inspectors track threats by examining the power production sites for enriched uranium required for generating the power for bombs, rather than that required for domestic use. Russia's Vladimir Putin likely recognized the economic and military value of nuclear power in Ukraine, and the dangers of a nuclear catastrophe by bomb or accident, when Russia invaded that country in February 2022.[7]

Places impacted by nuclear power are likely more numerous than we choose to acknowledge. Environmental regulations require utilities to cool and store, reprocess, or bury the waste byproduct, which can remain toxic for millennia.[8] The government effort that first enlisted nuclear energy to develop atomic bombs, known as the Manhattan Project, sourced most of its uranium ore from the Congo, Canada, and from the Colorado Plateau, specifically the Uravan Mineral Belt in Colorado and the Grants Mining District in New Mexico. Uranium ore is a naturally occurring metal embedded in the rocks of the earth's crust and found on many Indian reservations in the southwestern United States.[9] Uranium mining was central to the industrial and economic development of the Colorado Plateau.

A small percentage of naturally occurring uranium is in the form of the isotope U-235, which can undergo a process called nuclear fission. If a neutron strikes the naturally unstable atomic nucleus, the nucleus can split apart, yielding a large amount of energy. Such a fission also produces more neutrons, each of which can, in turn, cause a chain reaction. As found in nature, the concentration of U-235 is too small to sustain such a chain reaction, but a chain reaction becomes possible if uranium is purified to a higher U-235 concentration. The uranium can then serve as a fuel to generate electricity, or (if the uranium is purified to a still higher U-235 concentration) in an atomic weapon.

After mining the uranium, acid, leaching, filtering, and drying all process the heavy metal to create uranium oxide concentrate, often referred to as "yellowcake." Further enrichment creates a uranium dioxide powder, which when pressed becomes hard enough to form fuel pellets. Thin metal tubes encase these pellets, and they become the fuel rods that make up part of a nuclear reactor's core. The reactor uses these pellets for nuclear fission, which can generate very large amounts of low carbon or carbon-free electricity, using the heat generated in the nuclear reactor to turn water into a gaseous state (steam) that can spin a turbine. Nuclear waste, which remains toxic for millennia, necessitates containment. That waste and storage facilities are all part of the material culture of energy.[10]

In 1980, the United States passed environmental legislation, popularly known as Superfund, requiring the cleanup of long-term toxic sites. That process has necessitated more public transparency about environmental impact and consequences of energy extraction, including waste. Environmental and public health are important historical and current themes in nuclear energy. Furthermore, because nuclear power is a low carbon-producing energy and thus a potential alternative to fossil fuel energy, enhancing the public's understanding of its historical relationship to national security as both a weapon and for domestic use, can help us make better choices about including it in our energy future.

Histories and Contexts

Historians are still finding and interrogating historical sources that redefine energy's role in national security as both a military weapon and a domestic energy resource. They are asking new questions of new sources and sifting through declassified documents. Scientists continue to discover new advances in nuclear technology and gauge its past and future impact on people and the environment. The general public has large gaps in their understanding of nuclear power, from technical knowledge to environmental justice violations to issues of national security. Yet, nuclear energy has a sensitive and multifaceted history that includes all those impacted by nuclear power development and its use for both military and domestic purposes.

With all its tremendous power and a history associated with war, violence, tragedy, trauma, and toxicity, nuclear energy and its history are controversial. Much of the historical scholarship around nuclear energy specifically analyzes America's decision to use atomic bombs on the cities of Hiroshima and Nagasaki in Japan to end the Pacific conflict in World War II.[11] For years, most Americans have justified this military action to save American lives and quickly end a devastating war against a foe who showed no signs of surrendering. While Americans celebrated in the streets, the bombs devastated Japanese cities. The extent of this destruction is still highly contested, but most accounts (including the U.S. Department of Energy and the City of Hiroshima) conservatively estimate that the bombs killed between two hundred thousand to a quarter of a million people and obliterated over half of the buildings in each city. Most people died from burns from the blasts. Those who did not survived with radiation-caused health issues and psychological trauma for the remainder of their lives.[12]

Historical interpretation surrounding the first nuclear weapon remains highly contested in and out of the academy. Secretary of War Henry Stimson's explanation in 1946, published in the popular magazine *Harper's Weekly*, prevails in the collective memory. Stimson focused on assumptions that Japan would never surrender, and the atomic bomb was the best way to end the war and avoid more American casualties. As early as the 1960s, historians began questioning this narrative, citing documents that indicated that Japan was close to surrender. Instead, they argued, motives for the bombs included American efforts to display a show of technological and political power. Beyond these monographs, the Atomic Heritage Foundation and the Truman Presidential Library have strong resources that maintain the discussion and debates surrounding the decision to use the first atomic bombs. Alex Wellerstein's book *Restricted Data: The History of Nuclear Secrecy in the United States* (2021) and blog by the same name continue to raise questions and spirited debate around the history of nuclear secrecy.[13]

By the twenty-first century, scholars began to highlight the ecological and health consequences of uranium, mined at first to generate the nuclear power for the Manhattan Project. Through fieldwork, interviews, and examining early reports about uranium mining, Barbara Rose Johnston, Susan Dawson, and Gary Madsen have researched and documented evidence of environmental injustice resulting from uranium mining on Native lands, whose inhabitants often helped locate, mine, and mill the rocks. Tracy Voyles explored this impact even more fully in *Wastelanding* (2015), as did Sarah A. Fox in *Downwind* (2014) and the Navajo themselves in *The Navajo People and Uranium Mining* (2006). In response to litigation, Congress passed the 1990 Radiation Exposure Compensation Act to provide partial restitution to individuals for the effects of mining activity and the devastating health issues for Native miners and other exposed.[14]

Beyond the story of the Manhattan Project, scholarly interest in the history of nuclear energy has been limited but growing remarkably since the nuclear disaster at Fukushima, Japan, in 2011. For the industry itself, Martin Melosi's sweeping synthesis *Atomic Age America* (2013) serves as comprehensive and contextualized historical narrative that follows nuclear energy from the development of the atomic bomb through domestic power plants and into the twenty-first-century climate change debates.[15] Robert Lifset's edited volume can offer context for the 1970s environmental movement and oil crisis that briefly increased commercial energy interests in the nuclear power industry.[16]

In March 1979, a nuclear meltdown near Harrisburg, Pennsylvania, mirrored the fictional account of a nuclear accident in California portrayed in the film *The China Syndrome*, released fewer than two weeks before. A valve malfunction in the cooling system at the Three Mile Island power plant triggered a cascade of mistakes that caused a significant partial meltdown of the reactor core. The accident triggered a public previously nonplussed by nearby nuclear power plants to become terrified of the effects of radiation from them. The amount of radioactivity released is still debated among scientists, historians, and locals, but most agree that the accident scared the public so much that widespread fear and uncertainty permanently damaged the future of the nuclear power industry in America.

In the twenty-first century, several writers working in the public sphere chronicled nuclear accidents, including the Nuclear Regulatory Commission's historian Sam Walker's *Three Mile Island: A Nuclear Crisis in Historical Perspective* (2004), journalist Stephanie Cooke's *In Mortal Hands: A Cautionary History of the Nuclear Age* (2009), and research scientist James Mahaffey's *Atomic Accidents: A History of Nuclear Meltdowns and Disasters* (2014). Along with the others in Walker's series on regulating nuclear power, these works offer thorough and highly accessible overviews of the growth of the nuclear power industry.[17] Jacob Darwin Hamblin's *The Wretched Atom* provides a counternarrative about America exporting the promise of nuclear power on a global scale for geopolitical, as well as military purposes.[18]

While a larger radioactive release occurred on the lands of the Navajo reservation near the small village of Church Rock, New Mexico, a few months later, Three Mile Island (as the site of the most popularly known nuclear power accident) is an important historic resource in America's nuclear energy history beyond the Manhattan Project.[19] Yet this well-known site has little *public* interpretation save a single historical marker and a small, temporary exhibit at the National Museum of American History that now lives online that commemorated the twenty-fifth anniversary of the disaster.[20] Text explores the "Why?" and primarily blames the accident on human error, rather than the technology. Beyond a literal timeline, it does not address the accident's aftermath or the site's significance in the context of historic energy use and development, nor how it effectively halted future nuclear energy development in the United States.[21]

Nuclear sites have been problematic and controversial places for interpretation, particularly when trying to represent multiple perspectives. The role of technology in the politics of the Cold War, and the development of domestic energy in the same era, are historically parallel themes. Each history serves as context for the other. The following chapter reviews how the most prominent places affiliated with federal agencies and contractors have addressed the shift from nuclear power as a military weapon to a potential fuel source of the future.

Interpreting Sites of Nuclear Power

For years, preservationists have assessed nuclear sites as exceptions to the National Register's "fifty-year rule," referencing the amount of time normally required to achieve historical significance.[22] These sites are also exceptional for their association with military purposes. These interpretations, by and large, reflect the outsized role of the government in energy use, consumption, and research. Atomic and nuclear energy thus touches upon dozens of themes: military, policy, geopolitics, science, energy, labor, economy, environment, removal and land dispossession, health and safety, ethics, and alternative energy, depending on the locale and mission of the site. But public historians and activists have asserted that the interpretation of these sites has consistently lacked meaningful local community consultation or "civic engagement." They have called for a more complicated story of environmental and human consequences as a result of the development of nuclear bombs and energy.[23]

The 1995 cancellation of the *Enola Gay* exhibit at the National Air and Space Museum jolted historians in the academy and working in public-facing institutions. The very idea of reinterpreting history itself, not just the reasons for dropping the bomb, turned into a political firestorm and a flashpoint in the Culture Wars that thrived within the political partisanship of Bill Clinton's presidency and beyond. Those not previously embracing the values and practices of the only fifteen-years-young public history movement began to recognize the need for mediation between academic and public interpretation, particularly the importance of involving stakeholders in early planning stages through shared authority.

Figure 6.1. *Enola Gay*, Natonal Air and Space Museum, Steve F. Udvar Hazy Center, Smithsonian Institution, Chantilly, Virginia, 2014. Photo by author

Decades later, the *Enola Gay* controversy influences museum interpretation in general, but most notably has discouraged most places affected by nuclear energy from understanding that experience through a historical lens. Debates around military history overshadow the interpretation of the history of nuclear energy at sites that do discuss it. Both nuclear energy producers and the communities most affected by nuclear power development and use have perspectives to share. History museums located in these communities can and should help share these perspectives.

This history is complicated by the political origins of nuclear power development itself. Nuclear power development began as a government project, associated with a highly classified weapon. The federal government therefore owns and operates most of the sites and museums directly associated with nuclear power, and has maintained most of the control over the historical interpretation of nuclear energy for the public. Part of the reason for this involves federal and nuclear industry initiatives to build public support for nuclear power, once used as a military weapon, as a power we might enlist as an alternative domestic source of energy, even though the general public remains underinformed about its benefits and dangers.

The classified Manhattan Project intentionally fragmented its bomb-building operations into multiple sites, all working independently and unaware of their joint purpose. Connecting the sites and the resources involved through interpretation is one challenge in explaining this history within a technological and geopolitical context. Before and since the 1995 *Enola Gay* exhibit controversy, several federal museums focused on nuclear energy or energy in general, supporting the Atomic Energy Commission (established in 1946) in its task to expand nuclear power from military to civilian use. Advertisements and boosterism promoted nuclear power for its potential and jobs, assuring nearby residents of its safety and reliability. The most prominent government-directed museums are located in Albuquerque, New Mexico (National Museum of Nuclear Science and History, formerly the National Atomic Museum), Oak Ridge, Tennessee (American Museum of Science and Energy, originally the American Museum of Atomic Energy, established in 1949), and the Bradbury Science Museum in Los Alamos, New Mexico, founded in 1953 as the public face of the Los Alamos National Laboratory. Established in 1969 by congressional charter, the National Museum of Nuclear Science and Energy (affiliated with the Smithsonian), addresses the science and history of nuclear energy, emphasizing it as clean energy for the future. Still other sites of atomic development have confronted its controversial legacy, emphasizing efforts of cleanup and reconciliation.[24]

To move forward in interpreting nuclear energy, we need to absorb some lessons from the *Enola Gay* exhibit, namely attention to venue, timing, and early outreach efforts to incorporate multiple perspectives and stakeholders. Anthropologist Hugh Gusterson observed in 2006 that many of the federally affiliated museums focus on military history, and thus they are "seeking to present the development of nuclear weapons as natural facts of history rather than a continuing spur to controversy."[25] Gusterson implies that the goals and purposes of the federal agencies charged with operating government-owned or affiliated museums can compete with nuanced historical interpretation.[26] Perhaps only by addressing this difficult history can we start to reveal the connections and the differences inherent in using nuclear energy for military or domestic power purposes.

Just three years after the *Enola Gay* exhibit cancellation, former assistant secretary of energy and test site worker Troy E. Wade founded the Nevada Test Site Historical Foundation to participate in the "development and public exchange of views regarding the Nevada Test Site and its impact on the nation."[27] In close coordination with the Department of Energy (DOE) and embraced by the Smithsonian Institution, the foundation secured congressional funding and opened the Atomic Testing Museum (ATM) in 2005. Located just off the Las Vegas strip to entice tourists, the ATM dedicates eight thousand square feet to the testing in the desert that occurred seventy miles northwest of the museum from 1951 to 1992.

Despite the ATM's efforts, a sacred space to discuss deeply contested places and ideas remains elusive on the public history nuclear landscape. For some critics, the Las Vegas museum falls short of adequately addressing health and environmental effects, including community displacement, throughout its ten galleries and two theaters.[28] Simmering distrust about the history of the government's nuclear testing on Native lands has made it difficult to incorporate Native voices directly into the exhibits. Protesters, "downwinders," and Hiroshima/Nagasaki survivors have demanded more information and documentation at the museum regarding fallout and views of those connected to the site through contamination.[29]

The establishment of the multisite Manhattan Project National Historical Park (MPNHP) in 2014 came twenty years after the Smithsonian's *Enola Gay* exhibit debacle. When Cynthia Kelly came from the Environmental Protection Agency (EPA) to the DOE (the descendant agency of the Atomic Energy Commission) to work on cleaning up the former nuclear weapons complexes under its management, she became alarmed that the DOE had plans to raze all the project properties at Los Alamos as part of the CERCLA (Comprehensive Environmental Response, Compensation, and Liability Act), also known as Superfund. She left the DOE to try and preserve the atomic era resources. Kelly, who studied and taught history prior to her government service, founded the Washington, DC–based nonprofit Atomic Heritage Foundation (AHF) in 2002 to preserve and to interpret the remains of the Manhattan Project and its legacy.[30]

Based on an AHF request in 2004, Congress funded a special resource study to investigate the eligibility of a national park around what remained of the Manhattan Project. The study recommended three sites to make up a new park: Hanford, Washington; Los Alamos, New Mexico; and Oak Ridge, Tennessee. A congressional statute assigned the DOE to continue to own and manage the included sites, but the National Park Service (NPS) would take over interpretation through a Memorandum of Agreement, signed in 2015. Funded through government and private grant money, the AHF also supports the NPS's interpretive efforts and coordinates those of local stakeholders like the Los Alamos Historical Society, the B Reactor Museum Association, and the Oak Ridge Heritage Preservation Association to preserve the history of the Manhattan Project as well as "connecting the dots to innovations going on now in our national laboratories."[31]

The NPS convened a Scholars' Forum November 9–10, 2015, to begin developing interpretive themes, three years after recommendations from the landmark report *Imperiled Promise*, which encouraged the federal agency to engage in less instruction and rather to facilitate dialogue, develop empathy, engage new perspectives, and address difficult histories.[32] The forum coalesced around several broad categories consistent with previous sites, but also included other themes that expanded beyond the "triumphant" narrative.[33] John Hunner of New Mexico State University also hoped to include "the other uses of atomic energy from electrical production to nuclear medicine and the cultural and social transformations caused by the Manhattan Project."[34]

The interpretive themes that evolved from the forum included: the story of the Manhattan Project; the choices and consequences that created such a destructive force; sacrifice and displacement; revolutionary science and engineering and the impact on scientific advancement in technology and energy; and lastly, damage to humans and our environment. While these are tremendous interpretive opportunities, the MPNHP is still in the development stages, and it is unclear how much the park will explore the development and use of nuclear energy beyond the limited framework of the Manhattan Project.[35] However, NPS is working to expand interpretation in engaging ways that go beyond the traditional narratives, military perspectives, and scientific discoveries. To mark the seventy-fifth anniversary of the bombings, NPS launched a series of programs involving the Hiroshima Peace Museum called "Days of Peace and Remembrance" that included a social media campaign that shared photographs of postwar Hiroshima. NPS facilitated partnerships with the survivors of Hiroshima and Nagasaki in Japan. In 2021, NPS also began stakeholder meetings with African American and Native communities located near park sites, notably those whose members were impacted with adverse health effects from radiation.[36] One resource in particular bridges the history of the nuclear weapon to one of nuclear energy. While it is a unique resource, it is an example of the challenges and opportunities for resources associated with nuclear energy and power production.

CASE STUDY: The B Reactor, Hanford, Washington

The MPNHP includes the B Reactor, a gigantic, well-preserved artifact of the Manhattan Project used to produce nuclear energy. The B Reactor serves as both a historic site and a museum, and the DOE opened it for tours led by former employees at the site organized as the B Reactor Museum Association, who had worked to preserve the resource as a National Historic Landmark. By most accounts, the site is a unique one that offers visitors a place-based historic experience not unlike the coal mining museums discussed earlier, while communicating energy production in a way few sites can.

Most people appreciate the significance of the resource. Others, including journalists and activists, have remained troubled by its interpretive approach.[37] Historian of science Linda Richards first toured the site in 2009 and reviewed the B Reactor in 2016. While docents accessibly explain the history of nuclear science, she found the interpretation short on critical inquiry and shared authority, particularly in its lack of attention to the human and environmental impact of the atomic bomb, and transparency over safety conditions, past and present.

> Up close, the foreboding decrepit factory does not look like a museum. Inside, a corridor leads to the actual three-story reactor of graphite blocks, tubes, and valves. There, a DOE docent articulately explained exactly how the reactor worked: the way the fuel was inserted, the neutron reactions that changed uranium into plutonium, and the volume of Columbia River water that soared through it every minute. Between lectures, one is free to look through the building to see the decontaminated work and industrial spaces, from offices to break rooms to huge equipment and pipes. Videos show processes such as fuel discharge and processing and ventilation systems. . . . Wandering from room to room, visitors see the banality: 1940s-style phones, typewriters, desks, and other furnishings. Union pamphlets are on bulletin boards with safety and security reminders. Gas masks and safety equipment hang from hooks, and cubbies store Geiger counters. There are rows and panels of wires and large unidentified machines and artifacts. There is sparse-to-no textual interpretation: mundane equipment must speak for itself. Docents answered questions as if it were literally 1944, before any bombs were dropped—a strategic choice to show the feat of making plutonium, not actually using it.[38]

Richards and other critics continue to question the B Reactor, and other nuclear power sites, as a physically "safe space" for visitors due to contamination and radiation, or even as a place where they can freely engage in critical and thorough historical inquiry. However, she offered an update for this volume about recent programming that reaches out to additional stakeholders. Such outreach can move interpretation in a direction that acknowledges the dangers of nuclear energy development to workers, downwinders, victims, and possibly visitors.

NPS hosted their first official 75th commemoration of Hiroshima and Nagasaki in a virtual space in 2020. Perhaps the atomic bomb survivors' shared stories and art posted at the MPNHP commemoration made a sanctuary making visible more of the tragedy residing in every step of the global nuclear industrial complex from mining to production to use to waste. The next year, the NPS organized a "Stakeholder Engagement Project 2021.[39]

These well-intentioned focus sessions, representing multiple perspectives, were distressing. For example, facilitators sometimes underestimated the power differential between proponents of nuclear technology and the unacknowledged victims of it in the same virtual space. But the dissonance seemed worth it. In July, NPS heard from Hiroshima and Nagasaki first-, second-, and third-generation survivors and an atomic veteran. Why, asked Nobel Peace Prize winner Setsuko Thurlow, is there a celebration for fission but not the achievements of nuclear survivors for the UN Treaty on the Prohibition of Nuclear Weapons? Their courage and resilience to transform suffering into manifesting a nuclear weapons free future is a part of the story, too.[40] The group reiterated that there are vast repositories to access their vital testimony already in archives, oral histories, books, films, poetry and artworks. Ignoring them reinscribes harm.

NPS vowed to restructure and integrate what they have learned thus far into MPNHP's foundational document. The August 2021 MPNHP Hiroshima and Nagasaki commemoration held at Richland Park on the edge of the Columbia River featured luminary candles and mournful music, but it also included two third-generation Hiroshima survivors. One sang a stunning solo and shared some of her family's story, while the other was artist Yukiyo Kawano. She explained how her mother, Kuniko, was born after her grandparents survived Hiroshima. Being a third-generation Hibakusha "means it was my daily routine growing up to check with my mother and see how she was feeling each morning. It means I lost her to cancer in 2013—67 years after the U.S. Atomic bomb." She described her art "Suspended Moment" (sponsored by the New York Foundation of the Arts) because she was not allowed to display it at the Richland event. "I have been making life-sized recreations of Little Boy and Fat Man" Kawano said, "by stitching together inherited kimonos with my hair. In making the piece, I imagine weaving the past and present, the human-caused disaster that changed our world forever, and yet we continue 'examining what is human.' Because that is the only way that protects us from the future."[41]

Richards expressed hopefulness that the National Park Service would be more inclusive and transparent about issues of radiation. She sees progress and promise in future plans to install exhibits in two under-utilized rooms inside the reactor.[42]

While still limited by political sensitivities and classified information, federal agencies and nonprofit museums and organizations are cautiously beginning to address a broader people- and place-centered perspective. The experiences of energy technology and those who operated and developed that technology are incredibly important to interpreting energy. But there are other questions: Who and what occupied these places prior to the Manhattan Project? What changes did nuclear power promise or bring to communities after World War II? Economically? Environmentally? Through modernization? There are ready partners for such inquiries.

At the 2020 Western History Association conference, originally planned as an in-person event in Albuquerque, Rebecca Ulrich from the Sandia National Laboratories and Alan Carr from the Bradbury Museum in New Mexico invited lawyer Trisha Pritikin, a downwinder from Hanford who has organized the Consequences of Radiation Exposure Museum and Archives (CORE) in Seattle, Washington, to join a session about the commemoration of the use of the atomic bomb. Pritikin, who authored a book about lived experiences of radiation exposure from plutonium production at Hanford, later joined a panel with other downwinders, including Native people, to discuss the anniversary.[43] CORE's mission is to infuse the story of nuclear history with detailed accounts of the health effects of ionizing radiation from "uranium mining, milling or transport, nuclear weapons production, testing or use in warfare, nuclear reactor offsite releases or exposures" by preserving and amplifying the voices of victim populations.[44] The museum is part of the International Coalition of Sites of Conscience that advocates issues around civil and human rights, as well as environmental justice. CORE's mission was conveyed when a survivor of "Fat Man" visited the B Reactor as part of a Nagasaki Hanford Bridge Project Program sponsored by CORE and the city of Nagasaki in 2018. He admitted that there "are many explanations here, of what the process working here was like. But I'm left with a sense of what cruel things human beings do. . . . There was nothing . . . nothing about the suffering."[45]

A topic like nuclear power demands additional outreach and reconciliation to communities affected by nuclear testing and fallout, uranium mining, radiation, and nuclear waste storage. The Los Alamos History Museum established a project with Japan in 2016 "to develop dialogue with international museum colleagues and pursue understanding between Los Alamos, Hiroshima, and Nagasaki."[46] In 2018, the AHF published a guide for museums that address the atomic age, which includes Japanese museums and memorials to victims. They partnered with the National Museum of Nuclear Science and History in Albuquerque in the summer of 2019 to include "voices from Japan" through oral histories and interpretive vignettes addressing the legacy of nuclear power after World War II. One exhibit also examines nuclear waste transportation and the role of nuclear energy in future energy needs and in the field of medicine.[47]

CASE STUDY: The Atomic Legacy Cabin (The U.S. Department of Energy Office of Legacy Management), Grand Junction, Colorado

After a dozen or so years of heavy lobbying to acknowledge the impact of federal nuclear programs, primarily on the part of the Cold War and the defense industry, the DOE established the Office of Legacy Management (LM) in 2003. LM "is responsible for ensuring that DOE's post-closure responsibilities are met and for providing DOE programs for long-term surveillance and maintenance, records management, workforce restructuring and benefits continuity, property management, land use planning, and community assistance."[48] The office manages three still-developing interpretive centers at sites that the government used to refine uranium as part of the process of nuclear energy production. In 2007, historian Jason Krupar reviewed the development of two former uranium refinery sites, but he expressed disappointment that neither interpretive center employed historical context or best practices for enlisting the community in planning multiuse facilities.[49]

Figure 6.2.   Exterior, The U.S. Department of Energy Office of Legacy Management Atomic Legacy Cabin, Grand Junction, Colorado. Legacy Management, Department of Energy

Because most of the original buildings at all these sites required demolition for environmental reasons, and with the community active in preserving the site, LM built two new centers using new "green" construction. Fernald Preserve Visitor Center in Ohio (originally established in 2001 and redesigned in 2008) and Weldon Spring Site Interpretive Center in Missouri (originally established in 2003 with an opening of the newly designed center in 2022) offer vast spaces to interpret the remaining landscape which nature has reclaimed and thus, without interpretation, potentially obscures a controversial and contested history.[50] As LM expands historical interpretation at its facilities, it could offer a model for acknowledging complicated histories to educate and remember the human and environmental effects of energy development by working with the communities affected. The most recent interpretive center houses a robust historical exhibit that is a fraction of the size of those modern interpretive structures at Weldon Spring and Fernald.

The Atomic Legacy Cabin (ALC) is the most recent of LM's interpretive centers.[51] Because the Grand Junction, Colorado, site originally served as the national headquarters for acquiring domestic sources of uranium, the ALC focuses more on nuclear energy production than the other Manhattan Project sites. Since 1943, the DOE and its predecessor agencies (U.S. Army Corps of Engineers' Manhattan Engineer District, the U.S. Atomic Energy Commission, and the U.S. Energy Research and Development Association) have continuously operated at the site to stage, inform, and collect aspects of federal and private uranium exploration. During World War II, fewer area residents had any awareness of the region's historical role in nuclear power development than during the Cold War. LM still monitors the site for groundwater quality. The original cabin, still relatively isolated on the Colorado Plateau, but with railroad access and a reliable source of water, maintains its original building envelope or the exterior.[52]

Predating the Manhattan Project, the building is the oldest on the site, and itself a historic artifact. It served as the office of the site's first manager, Phillip Leahy, and it also coordinated much of the activity of the Colorado Plateau. Exterior signage emphasizes the Manhattan Project, mining, and LM's mission of public communication. This small museum, which opened to the public on June 6, 2019, provides a key origin piece of the nuclear energy production story missing at other place-based sites. The interpretive focus still tells the story of the Manhattan Project through uranium, its mining and development on the Colorado Plateau, and the people involved in developing it. According to staff, "The early stages of the uranium production cycle often do not get as much attention as the later stages. . . . Mining and milling of uranium are the necessary *first* steps toward building a nuclear device or generating nuclear energy."[53]

The interior interpretation synthesizes a great deal of historical, geological, and technological content into the 949-square-foot exhibit space of the original cabin. Listing the property on the National Register of Historic Places in 2016 seems to have encouraged richer historical interpretation than at other LM sites. While noted for its role in the development of the atomic bomb, broader interpretive themes in the cabin's exhibit space help contextualize this place within scientific, medical, cultural, economic, and political contexts of World War II and the Cold War. The cabin also served a significant role with the Atomic Energy Commission (AEC), which Congress established in 1946. The civilian-controlled AEC's primary goal was to develop, use, and control atomic energy for military and civilian applications. The agency promoted uranium exploration, mining, and processing efforts throughout the nation, particularly on the Colorado Plateau due to its geology, geography, and topography. The exhibit text and images explain the presence of uranium, uranium mining, and uranium from "ore to green sludge to yellow cake." Exhibits address the local economic and environmental impact of mining, and LM's environmental remediation efforts to clean up waste at the site. Indeed, "a driving force behind the creation of the ALC was to help explain why LM looks after disposal cells containing radiological contamination in the Four Corners region."[54]

LM added staff people with formal training in history, and it built upon its experiences at Weldon Spring and Fernald by partnering with its stakeholders to renovate the cabin as an interpretive center, including exhibit space and a classroom. Designers opened the cabin's interior space for the exhibit by removing interior walls, but many architectural details and furnishings reflect the time period. Museum professionals, specifically from the Museums of Western Colorado, an American Alliance of Museums-accredited, local history-focused organization based in Grand Junction, helped to incorporate stories of local interest. These included stories about the radium miner Pegleg Foster and Calamity Camp—a regional mining area that provided radium and later uranium.[55] The exhibit also introduces terminology and historical figures related to the science of atomic energy (i.e., Marie Curie), the participants of the Manhattan Project, and residents of the plateau, including the Navajo Code Talkers, since "they were people from the Colorado Plateau who fought in the Pacific Theater, which was where the first atomic bombs were deployed."[56]

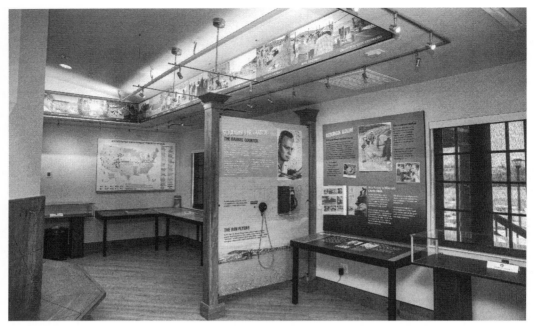

Figure 6.3.   Interior Exhibit of ALC, Grand Junction, Colorado, 2021. Legacy Management, Department of Energy

Exhibit text notes that due to the secrecy surrounding the Manhattan Project, most of the workers and military in Grand Junction, including the commanding officer, did not realize their purpose until the bombings of Hiroshima and Nagasaki. While it does not address ethical or political issues around the development and use of the atomic bomb, the exhibit includes casualty statistics and photographic evidence documenting the destroyed buildings in both Japanese cities that were civilian in nature (the former exhibition hall/now the Hiroshima memorial, and a cathedral). Text stops short of exploring the trauma and devastation left in Japan, but it addresses the search for and mining of uranium on local Native lands, including LM's long-term stewardship programs with tribes like the Navajo Nation today.[57]

While not strictly interpretive, the ALC and LM's programming and outreach might help develop a local consumer base and workforce to transition nuclear energy from a weapon of war to peacetime use. Since opening just prior to the COVID pandemic, the cabin has reached thousands of new visitors both in person and virtually. Like Fernald and Weldon Spring, the ALC provides science, technology, engineering, and math (STEM) programming, offered on-site, off-site, or virtually in partnership with local and regional STEM initiatives and groups. STEM programs at local high schools and colleges on the Navajo Nation and Hopi reservations specifically focus on the landscape of the Navajo Nation and relate the LM's work at nearby disposal cells. Outreach includes presentations at local chapter houses (local tribal meeting spaces) and other community spaces, participation in large signature events on the Navajo Nation, and involvement in the annual American Indian Science and Engineering Society conference.

Artifact Spotlight

Nuclear Landscapes

Nuclear energy's dramatic impact on a place can be dramatic and instructive, and these landscapes become artifacts of energy history. Interpretation can help visitors read the energy history upon the landscapes. John Sprinkle noted atomic sites as "of exceptional importance" and deserving of interpretive and preservation attention.[58] John Wills wrote about the resilience of the landscape—something that the Office of Legacy Management emphasizes at Weldon Spring and Fernald to supplement narratives of ruin and horror. Government officials chose these landscapes precisely because of their remote location and "wild" landscape. He explains that "Hanford Engineering Works and Nevada Test Site represented sacrifice zones, Armageddon wastelands where humans experimented with deadly materials" and that "nuclear lands were inescapably tied to partisan interpretations of the nuclear age."[59]

Without exploring the cultural and environmental history of former nuclear sites, many now hidden by ecological resilience, visitors can erroneously read these landscapes as "natural parks." To understand these places as experimental landscapes, historian Andy Kirk examined declassified photographs taken for the Civil Defense Administration that captured the technological marvel of the detonations but not the environmental context of

Figure 6.4. Two-Story Wood-frame House, Nevada, 1955. Records of the Office of Civil and Defense Mobilization (RG 304), National Archives

the desert. Under the bright sunlight of midday, these images washed out detail to emphasize a sense of empty places needing to be populated by mannequins. Such documentation enforced the notion of atomic energy as safe and contained. They "shaped the public perception about the testing region" as an empty place absent of nature or history that would avoid questions over nuclear waste disposal, the existence of downwinders and thoughtful, consequential, and ethical discussions about nuclear weapons and power.[60]

Exhibits and interpretive programs can also take place in community spaces and art museums rather than historical sites or museums. Nuanced understanding about these complicated cultural landscapes may be hard to provoke, but photographer Julian Kilker suggests that enlisting low-light photography of these places can convey fragility in ways narratives cannot or even casual in-person viewing might not catch. By altering initial perceptions of these landscapes, these images can encourage more dialogue about how nuclear energy development, whether for weapons or for electrical use, affects humans and our environment. Glenna Cole Allee and Abbey Hepner have similarly documented the nuclear industry, with the goal of making the invisible visible, through photography. With a downwinder providing the text, Hepner's work links itself to the material properties of the subject through x-rays and uranotypes (uranium prints).[61]

Local and University Collections

The public historical interpretation of nuclear energy development has remained fairly static in museums, but scholars and local institutions like the Columbia River Exhibition of History, Science and Technology Museum, founded in 1962 as the Hanford Science Center, have actively collected artifacts and documented local perspectives around nuclear history beyond World War II and the Cold War. When that institution closed in 2014, the University of Washington Tri-Cities in Richland absorbed its collections and built upon them, additionally acquiring, preserving, and curating the DOE's Hanford collection in what they have named the Hanford History Project. Robert Bauman and Robert Franklin publish a series on Hanford's nuclear history and loan out artifacts like government reports, photographs, and newspaper and newsletter articles. They also collect and curate tangential oral histories and local donations.[62]

Decorative Housewares and Kitsch

The Oak Ridge Associated Universities consortium holds examples of housewares, including other widely distributed items like comics, signs, soft drink bottles, and artifacts using atomic imagery that many museums might find in their collections and that we now often characterize as "kitsch." There is something unsettling about including objects of kitsch related to "atomic culture" in places like Richland and juxtaposing that against the radiation effects on Nagasaki. But that unsettling discomfort might just be an effective and accurate way to connect to visitors.

The Atomic Legacy Cabin enlists objects to both contextualize and illustrate its central object of inquiry: uranium. Uranium is the essential fuel for causing a chemical reaction for producing nuclear energy, but like petroleum, it is an energy-related resource found in many

Figure 6.5.   Glassware display. The Barnes Museum, Southington, Connecticut, 2022. Photo by Molly Norton

Figure 6.6.   Uranium Glass detail. The Barnes Museum, Southington, Connecticut, 2022. Photo by Molly Norton

forms due to its unique behavior. The exhibit features a radio and fireside chat by President Roosevelt, but one display of Vaseline glass and a discussion of Fiestaware shows uranium's use in dinnerware. Coated with uranium oxide to achieve certain colors, red Fiestaware plates and yellow green hued "Vaseline" glass, which glows under a black light, help bring this mineral both literally and psychologically into one's everyday world.

Interpreting these pieces, found mixed among museum houseware collections across the country, can help visitors to reflect upon the *chemical* nature of energy resources and their presence in the kinds of materials we use as our weapons of warfare but also in our household objects and what powers our appliances. Exhibit text should also convey that while not considered dangerous, some of these items may still retain at least small amounts of radioactivity.[63] What does and does not contain radium (created from the decay of uranium atoms) is sometimes confusing since advertisers purposely tried to fool consumers. Albuquerque's National Museum of Nuclear Science and History possesses materials that advertised radium but do not contain it. In February 2020, a curator at the Atomic Testing Museum discovered a still-radioactive artifact (a radiendocrinator that boasted dubious medical benefits when worn against the skin), which is now buried in the desert with other radioactive waste. There are likely still many more such artifacts in museum collections and available at public auction. Educational materials should guide curators and consumers in the proper care and safe storage of, and if needed, disposal of these artifacts.[64]

Interpreting nuclear energy must not only address the development of a nuclear bomb, but how nuclear power has evolved domestically. Both intertwined narratives must include more conversations about environmental remediation, as well as environmental justice issues surrounding waste storage, and they must assume some of the hard work of reconciliation with communities affected by nuclear testing and fallout, uranium mining, nuclear power development, and production for domestic use.[65] Due to ongoing ethical questions and complicated issues around safety, environmental, and health concerns, the interpretation of nuclear power demands multiple voices and perspectives. Strong partnerships between government agencies, utilities, advocacy groups, and local historical and community organizations can provide multiple perspectives to engage public conversations across the many communities, sites, and landscapes touched by nuclear power development and generation. Public art projects, like Navajo artist Ed Singer's "Dear Downwinder," and Yukiyo Kawano's installations, record often absent perspectives on nuclear power.[66] The 2017 *Hope and Trauma in a Poisoned Land* exhibit at the Coconino Center for the Arts in Flagstaff, Arizona, featured twenty local Native and non-Native artists commenting on the impact of uranium mining on the Navajo Nation in response to a four-day educational program from community members, scientists, and health officials.[67]

Few museums, even those affected by nuclear power or that are located in communities that produce and use nuclear power, attempt to critically examine nuclear energy history regarding domestic power production. The history of nuclear energy in public memory, beyond the triumphant narratives of military victory and technological feats, is often shrouded in fear, warfare, injustice, and danger. However, this problematic history offers important context as the world searches for fossil fuel alternatives to reduce carbon emissions while meeting modern energy demands.

Notes

1. Mark O'Connell, "Why Would Anyone Want to Visit Chernobyl? Maybe they're looking for a glimpse of the Apocalypse," *New York Times*, March 24, 2020.
2. J. John Lennon and Malcolm Foley, *Dark Tourism: The Attraction of Death and Disaster* (London: Thompson, 2000); George Johnson, "The Nuclear Tourist: An Unforeseen Legacy of the Chernobyl Meltdown," *National Geographic*, October 2014; Jenna Berger, "Nuclear Tourism and the Manhattan Project," University of Houston, submission for the *Columbia Journal of American Studies*, 2004, http://www.columbia.edu/cu/cjas/print/nuclear_tourism .pdf?q=manhattan-project-test-site.
3. Jessica Moody, "Where Is 'Dark Public History'? A Scholarly Turn to the Dark Side, and What It Means for Public Historians," *Public Historian* 38:3 (August 2018), 114; Rebecca Boyle, "Greetings from Isotopia: Why would anyone visit a radioactive ghost town or the remnants of a nuclear reactor?" *Distillations: Science History Institute* (October 12, 2017).

4. Matt Wray, "A Blast from the Past: Preserving and Interpreting the Atomic Age," *American Quarterly* 58:2 (June 2006), 468.

5. Accident risk is increasing, as many nuclear power plants and facilities (about 441 current plants in thirty countries) are at the edge of their intended and licensed lives, and experimental versions are under construction. Dave Lochbaum, "All Things Nuclear: Nuclear Bathtub Safety," September 13, 2016, Union of Concerned Scientists, accessed January 21. 2022, https://allthingsnuclear.org/dlochbaum/nuclear-bathtub-safety.

6. Dr. Egle Rindzeviciute, "NuSpaces-Nuclear Spaces: Communities, Materialities, and Locations of Nuclear Cultural Heritage," Heritage Research Hub, accessed January 21, 2022, https://www.heritageresearch-hub.eu/project/nuspaces.

7. See World Nuclear Association, "Ukraine: Russia-Ukraine War and Nuclear Energy," https://world-nuclear.org/information-library/country-profiles/countries-t-z/ukraine-russia-war-and-nuclear-energy.aspx, accessed May 10, 2022.

8. Samuel J. Walker, *Three Mile Island: A Nuclear Crisis in Historical Perspective* (Berkeley: University of California Press, 2006), 51.

9. Outside of the United States, uranium is also found and mined in Kazakhstan, Canada, Australia, Namibia, Niger, and Russia.

10. Martin Melosi, *Atomic Age America* (Routledge, 2012); US Energy Information Administration, https://www.eia.gov/energyexplained/nuclear, accessed November 2, 2020.

11. See bibliography.

12. "Hirsohima and Nagasaki: 75th Anniversary of the Atomic Bombings," BBC News, August 9, 2020, https://www.bbc.com/news/in-pictures-53648572, accessed December 22. 2021; Atomic Heritage Foundation, "Hiroshima and Nagasaki Bombing Timeline," https://www.atomicheritage.org/history/hiroshima-and-nagasaki-bombing-timeline, accessed December 22, 2021; Atomic Archive, "The Atomic Bombings of Hiroshima and Nagasaki," accessed December 22, 2021.

13. Sam J. Walker, "The Decision to Use the Bomb: A Historiographical Debate," *Diplomatic History* 14:1 (Winter 1990), 97–114; Gar Alperovitz, Richard Rhodes, and Martin Sherwin, *A World Destroyed: Hiroshima and the Origins of the Arms Race* (New York: Vintage Books, 1987 reprint), xxii; Alex Wellerstein, blog, http://blog.nuclearsecrecy.com/2015/08/03/were-there-alternatives-to-the-atomic-bombings; Michael Gordin, *Five Days in August: How World War II Became a Nuclear War* (Princeton, NJ: Princeton University Press, 2007); Alex Wellerstein, *Restricted Data: The History of Nuclear Secrecy in the United States* (Chicago: University of Chicago Press, 2021); Edward T. Linenthal, "Anatomy of a Controversy," in *History Wars: The Enola Gay and Other Battles for the American Past*, eds. Edward T. Linenthal and Tom Engelhardt (New York: Metropolitan Books, 1996), 52.

14. Johnston, Dawson, and Madsen, "Uranium Mining and Milling," 111–34; Traci Brynn Voyles, *Wastelanding: Legacies of Uranium Mining in Navajo Country* (Minneapolis: University of Minnesota Press, 2015); Sarah Fox, *Downwind: A People's History of the Nuclear West* (Lincoln, NE: Bison Books, 2014); Doug Brugge, Esther Yazzie-Lewis, Timothy H. Benally, eds., *The Navajo People and Uranium Mining* (Albuquerque: University of New Mexico Press, 2007).

15. Martin Melosi, *Atomic Age America* (New York: Routledge, 2012).

16. Robert Lifset, ed., *American Energy Policy in the 1970s* (Norman: University of Oklahoma Press, 2014).

17. Melosi, *Atomic Age America*; James Mahaffey, *Atomic Accidents: A History of Nuclear Meltdowns and Disasters* (New York: Pegasus Books, 2014); Walker, *Three Mile Island*; Righter, 149–69.

18. Jacob Darwin Hamblin, *The Wretched Atom: America's Global Gamble with Peaceful Nuclear Technology* (New York: Oxford University Press, 2021).

19. Linda Richards, "On Poisoned Ground," *Distillations: Science History Institute* (April 21, 2013); Doug Brugge, Jamie L. deLemos, and Cat Bui, "The Sequoyah Corporation Fuels Release and the Church Rock Spill: Unpublicized Nuclear Releases in American Indian Communities," *American Journal of Public Health* 97:9 (September 2007), 1595–600.

20. Smithsonian National Museum of American History, "Three Mile Island: The Inside Story," https://americanhistory.si.edu/tmi/tmi01.htm, 2004, now online, accessed September 15, 2021.

21. Paul Forman and Roger Sherman, "Three Mile Island: The Inside Story," Smithsonian Museum of American History, https://americanhistory.si.edu/tmi, accessed June 30, 2021; When the Pennsylvania Historical and Museum Commission launched a 150th commemoration of the Drake oil well in 2009, nuclear power remained absent from the state review of energy history. However, the State Museum of Pennsylvania added a Remote Reconnaissance Vehicle, the type of machine involved in the cleanup, that year to the Hall of Industry and Technology (donated in 2000 by the operator at the time of the accident, which served as part of the cleanup—at least to train operators on). Carnegie Mellon University, "Remembering Three Mile Island," https://www.cmu.edu/homepage/environment/2009/spring/30-years-ago.shtml, accessed September 21, 2021.

22. John H. Sprinkle Jr. "'Of Exceptional Importance': The Origins of the 50-Year Rule in Historic Preservation," *Public Historian* 29:2 (Spring 2007), 94–96.

23. Jason Krupar, "Burying Atomic History: The Mound Builders of Fernald and Weldon Spring," *Public Historian* 29:1 (Winter 2007), 32–33; Paul Williams, "Going Critical: On the Historic Preservation of the World's First Nuclear Reactor," *Future Anterior: Journal of Historic Preservation, History, Theory, and Criticism* 5:2 (Winter 2008), vi, 1–18; Arthur Molella, "Atomic Museums of (Partial) Memory," *Journal of Museum Education* 29:2/3, *Museums of Memory* (Spring/Summer–Fall 2004), 21–25; W. Patrick McCray, "Viewing America's Bomb Culture: The Atomic Testing Museum. Las Vegas, Nevada," *Public Historian* 28:1 (2006): 152–55; Linda Marie Richards, "The B Reactor NHL. The Hanford Site. Manhattan Project National Historical Park (409th)," *Public Historian* 38:4 (November 2016), 305–17.

24. Office of Legacy Management, "Exhibits, Museums, and Historic Facilities," accessed January 1, 2022, https://www.energy.gov/lm/doe-history/historical-resources/exhibits-museums-and-historic-facilities.

25. Hugh Gusterson, "Nuclear Tourism," *Journal of Cultural Research* 8:1 (May 2006), 24.

26. The federal government established the American Museum of Science and Energy, originally the American Museum of Atomic Energy, in 1949 to educate (and likely influence) the public about atomic energy and its uses. While initially focusing on the (quite celebratory) narrative of the Manhattan Project at Oak Ridge, the human toll of its victims in Japan or nearby communities was absent. In 1975, the museum moved to a newer fifty-five thousand-square-foot building and eventually expanded its mission to interpreting atomic energy in general, the 1970s oil crisis, other forms of renewable energy like solar, and STEM topics in general.

27. Wray, 469. Sociologist Matt Wray also suggested that the *Enola Gay* controversy, and how it was (not) resolved, should be a "teachable moment" about memory and history, but the ATM and other museums still default to technological achievements dominating the narrative, rather than reflecting the changing or evolving historiography, including the

"counternarratives/narratives of dissent" found in archives like the Nevada Test Site Oral History Project and the Nuclear Testing Archive.

28. In a review for the *Public Historian* in 2006, Patrick McCray noted that even using the word "atomic" over "nuclear" implies a focus on military and defense. He observed that the museum views the role of nuclear energy through Cold War politics, economy, and culture with traditional interpretation through artifacts of popular culture, and the perspectives of participants, soldiers, and spectators over victims. The impact of testing on the cultural or natural landscape is limited, as are voices who opposed testing, and those who lived "downwind" of the blasts and have since cited various health impacts. See McCray, "Viewing America's Bomb Culture," 152–55; Greg Mitchell, "The Folks Who Are Sanitizing and Celebrating the History of the Bomb," *History News Network*, 2005, https://historynewsnetwork.org/articles/11653.html.

29. Bryan C. Taylor, William J. Kinsella, Stephen P. Depoe, Maribeth S. Metzler, eds., *Nuclear Legacies: Communication, Controversy, and the U.S. Nuclear Weapons Production Complex* (Lanham, MD: Lexington Books, 2008).

30. At the time, the V-site and Experimental Breeder Reactor near Idaho Falls—a power plant that experimented with producing electricity from atomic energy and plutonium in 1951 for domestic commercial purposes—also met National Historic Landmark standards. The AHF was instrumental in establishing the park. This illustrates the attempt to transition from generating nuclear power for military purposes to generating nuclear energy for commercial and domestic purposes. Katherine Reece, "Profiles: Preservation and the Manhattan Project: Cynthia Churchill Kelly," *Wellesley Magazine* (Winter 2015), https://magazine.wellesley.edu/winter-2015/preservation-and-the-manhattan-project, accessed July 30, 2021.

31. "Preserving Manhattan Project History," AmericanHistoryTV, https://www.c-span.org/video/?507055-2/preserving-manhattan-project-history, November 12, 2020; "National Museum of Nuclear Science and History," https://www.manhattanprojectvoices.org, accessed September 27, 2021.

32. Anne Mitchell Whisnant, Marla R. Miller, Gary B. Nash, David Thelen, "Imperiled Promise: The State of History in The National Park Service," Organization of American Historians, 2011 (updated 2016).

33. National Park Service, "Scholar's Forum Report, November 9–10, 2015," Manhattan Project National Historical Park, New Mexico, Tennessee, Washington: National Park Service, United State Department of the Interior, 2015.

34. NPS, Scholar's Forum, 26–27. With support from several organizations and academic groups like the Organization of American Historians, affected community groups and supporters wrote a letter to the MPNHP project manager demanding that the newly established Manhattan Project National Historical Park incorporate downwinder stories, engage in consultation with affected communities, victims, and advocates (including Indigenous nations associated with the sites), feature the history of movements organized to protest nuclear power and weapons development, and reveal data on radiation releases through the Cold War in some part of the park interpretation. Letter from CORE to Tracy Atkins, project manager, Manhattan Project National Historical Park, Denver Service Center, August 27, 2015, https://www.corehanford.org, accessed July 17, 2021.

35. Scholar's Forum, 22; "Foundational Document," Manhattan Project National Historical Park, National Park Service, January 2017.

36. Kris Kirby, superintendent, Manhattan Project National Historical Park, to Leah S. Glaser, phone conversation, December 9, 2021.

37. Williams, "Going Critical," 2, 14. In a 2008 review, Williams felt the B Reactor left out critical themes "including the abject devastation of the impact on Hiroshima and Nagasaki, the development the American military-industrial complex, the political culture of the Cold War, the development of the peace movement, the long-term environmental effects of nuclear waste, and the place of nuclear weapons in today's politically uncertain times." He argued that its current preservation reflects a provincial, place-based perspective, separated from the killing of seventy-five thousand Nagasaki civilians it enabled, and urged that the preservation of this site allow visitors to connect to the critical stories and other places including Nagasaki, but also Three Mile Island and Chernobyl, something he calls "synchronous heritage."

38. Linda Marie Richards, "The B Reactor NHL. The Hanford Site. Manhattan Project National Historical Park (409th)," *Public Historian* 38:4 (November 2016), 305–17; Hanford Site website, "100 Area," http://www.hanford.gov/page.cfm/100Area; US Department of Energy, "Cleanup Progress at Hanford," June 2016, http://www.hanford.gov/c.cfm/media/attachments.cfm/DOE/Cleanup_Progress_at_Hanford_Factsheet-06-2016.pdf.

39. See Manhattan Project, National Park Service, https://www.nps.gov/mapr/getinvolved/stakeholder-engagement-project-2021.htm; National Park Service, "Manhattan Project National Historical Park Hosting Days of Peace and Remembrance in observance of atomic bombings of Japan," August 5, 2020, https://www.nps.gov/mapr/learn/news/manhattan-project-national-historical-park-hosting-days-of-peace-and-remembrance-in-observance-of-atomic-bombings-of-japan.htm; "75th Commemoration," https://www.nps.gov/mapr/learn/historyculture/75th-commemoration.htm, "Days of Peace and Remembrance, Lights for Peace," https://www.nps.gov/mapr/blogs/commemoration-with-lights-for-peace.htm, 2020; and "Messages of Peace," https://youtu.be/u5HM-bRKfFU 2021, accessed September 21, 2021.

40. Thurow, along with Beatrice Fihn, accepted the 2017 Nobel Peace Prize on behalf of ICAN, https://www.icanw.org/nobel_prize.

41. View her art and bio at http://yukiyokawano.com/about. Also see the organizational pages https://www.tridec.org, and Annette Cary "'Just plain wrong.' Tri-Cities leaders blast Washington State over Hanford nuclear waste rule," *Tri-city Herald*, March 5, 2021, https://www.tri-cityherald.com/news/local/hanford/article249681723.html.

42. Phone conversation with Superintendent Kris Kirby, December 9, 2021; phone conversation with Linda Marie Richards, January 21, 2022.

43. Western History Association Annual Conference Program, "Migrations, Meeting Grounds, and Memory" (October 13–17, 2021), 43, 55. Also see Trisha Pritikin, *The Hanford Plaintiffs: Voices from the Fight for Atomic Justice* (Lawrence: University Press of Kansas, 2020).

44. "Consequences of Radiation Exposure," https://www.corehanford.org, accessed July 17, 2021.

45. "Nagasaki Survivor Tours B-Reactor," *NBC News* (November 28, 2018); "Nagasaki Survivor Visits Hanford, Finds Some of the Story Still Untold," *Seattle Times* (updated March 12, 2018).

46. The Los Alamos Museum, https://www.losalamoshistory.org/losalamos_japan_project.html; "Museums and the Atomic Age," https://www.atomicheritage.org/history/museums-and-atomic-age, accessed July 18, 2021.

47. The National Museum of Nuclear Science and History, Albuquerque, New Mexico, https://www.nuclearmuseum.org/, accessed September 1, 2021. Alison Fields provides a fascinating and introspective overview of how museums have interpreted the atomic age in her essay, "Narratives of Progress and Peace: Atomic Museums in Japan and New Mexico," part of her

book, *Discordant Memories: Atomic Age Narratives and Visual Culture* (Norman: University of Oklahoma Press, 2020), 55–82.

48. "Our Promise of Protection," https://www.energy.gov/lm/about-us, accessed December 21, 2021.

49. Krupar, 41, 44–46; Jason Krupar and Stephen Depoe, "Cold War Triumphant: The Rhetorical Uses of History, Memory, and Heritage Preservation Within the Department of Energy's Nuclear Weapons Complex," in Bryan Taylor (eds.), *Nuclear Legacies: Communication, Controversy, and the U.S. Nuclear Weapons Complex* (Lanham, MD: Lexington Books, 2008), 135–56; Jason Krupar, "The Challenges of Preserving America's Nuclear Weapons Complex," in Rosemary B. Mariner and G. Kurt Piehler (eds.), *The Atomic Bomb and American Society: New Perspectives* (Knoxville: University of Tennessee Press, 2009), 381–406.

50. The Perspectives Group for the Fernald Citizens Advisory Board, "Telling the Story of Fernald: Community-Based Stewardship and Public Access to Information," October 2002, The Perspectives Group, https://www.theperspectivesgroup.com/links/Tellingstory.pdf, accessed September 20, 2021. The University of Cincinnati's Center for Environmental Communication Studies houses an oral history program designed to collect the remembrances of individuals who either worked or lived near Fernald from 1951 to the present. Fernald Community Alliance, http://www.fernaldcommunityalliance.org, accessed September 21, 2021; Krupar, "Burying Atomic History," 57. The Rocky Flats Cold War Museum is entirely online, commemorating the AEC facility in Colorado, but unlike Fernald and Weldon Spring, it has no interpretive center (despite congressional approval), and is now, like those places, a natural preserve. Rocky Flats Cold War Museum, http://www.rockyflatshistory.org/videos.html, (accessed September 20, 2021).

51. YouTube, "Atomic Legacy Cabin Virtual Tour- A Manhattan Project Site," August 18, 2021, https://www.youtube.com/watch?v=60G5hK7_ToI, accessed September 22, 2021.

52. Jonathon C. Horn, Alpine Archeological Consultants, "Department of Energy Grand Junction Office," National Register Nomination Form, approved June 10, 2016.

53. Office of Legacy Management staff (LM) correspondence to Glaser, July 19, 2021.

54. Photographs via email, LM to Glaser, June 17, 2021; LM to Glaser, July 19, 2021.

55. LM to Glaser, July 19, 2021.

56. LM to Glaser, July 19, 2021.

57. Atomic Legacy Cabin exhibit, Grand Junction, Colorado.

58. John Sprinkle, "Of Exceptional Importance: The Origins of the 50-Year Rule in Historic Preservation," *Public Historian* 29:2, 81–103.

59. John Wills, "'Welcome to the Atomic Park': American Nuclear landscapes and the 'Unnaturally Natural'," *Environment and History* 7:4 (November 2001), 465–66.

60. Andy Kirk, "Rereading the Nature of Atomic Doom Towns," *Environmental History* 17:3 (July 2012), 634–47.

61. Julian Kilker, "Visualizing What Happened Near Vegas: Experiences in Photographing a Public History Project," *Public Historian* 36:3 (August 2014), 100–08; Abbey Hepner with Kirsten Pai Buick, and Nancy Zastudil, *The Light at the End of History* (Durham, NC: Daylight Books, 2021); Glenna Cole Allee with Mark Auslander, *Hanford Reach: In the Atomic Field* (Durham, NC: Daylight Books, 2021).

62. Interview by author with Robert Bauman and Robert Franklin, Zoom interview, July 27, 2021. Also see Robert Bauman, "Teaching Hanford History in the Classroom and in the Field," *Public Historian* 29:4 (Fall 2007), 45–55; Robert Bauman and Robert Franklin, eds.,

Nowhere to Remember: Hanford, White Bluffs and Richland to 1943, Hanford History Series, Vol. 1 (Pullman: Washington State University Press, 2018): *Echoes of Exclusion and Resistance: Voices from the Hanford Region, Hanford History Series, Vol. 3* (Pullman: Washington State University Press, December 2020); Hanford History Project, Washington State University–Tricities, https://tricities.wsu.edu/hanfordhistory, accessed July 28, 2021.

63. Office of Legacy Management, Exhibit, Grand Junction, CO; Oak Ridge Associated Universities, Museum of Radiation and Radioactivity, "Fiestaware, ca. 1930s," and "Vaseline and Uranium Glass, ca. 1930s," https://www.orau.org/health-physics-museum/collection/consumer/ceramics/fiestaware.html, accessed September 26, 2021.

64. Randy Dotinga, "The Lethal Legacy of Early Twentieth Century Radiation Quackery," *Washington Post* (February 15, 2020); see R.W. Holmes, *Substance of the Sun: The Cultural History of Radium Medicines in America.* Dissertation. University of Texas at Austin. 2010.

65. Richard Rhodes, "Why We Should Preserve the Manhattan Project," *Bulletin of the Atomic Scientists* 63:3 (2007), 34–43.

66. Edward Singer, "Downwind," downwindhistory.com, accessed July 18, 2021; View Kawano's art and biography at http://yukiyokawano.com/about.

67. The National Endowment for the Humanities funded this collaboration between the Navajo Nation, the Flagstaff Arts Council, University of New Mexico Community Environmental Health Program, and Northern Arizona University.

CHAPTER 7

Emphasizing Energy Diversity

The Past and Future of Renewable Energy

REVIOUS CHAPTERS DISCUSSED how both private industry and government agencies have historically promoted and supported the production and use of coal, petroleum, natural gas, and nuclear power to meet energy needs. Historian Robert Righter argues that oil rigs and coal-fired steam plants became associated with economic progress to the detriment of less centralized, more independently operated, and essentially free renewable energy sources like wind.[1] The general public considers alternative and renewable energy sources and technologies in the context of politics, technology, and ecology, but rarely in the context of history. To the comparatively modest extent that history museums can and do educate the public about energy, interpretation is only beginning to encompass discussions of diversifying and transitioning. Even science museums present the history of alternative energies piecemeal or as prologue to the ever-distant future.

As cultural institutions, historic sites and museums are logical places to tackle the massive cultural shift our society needs to collectively make for our next energy transition. Renewable energies like solar, wind, geothermal, and biofuel require the same level of attention that museums and historic sites bestow on fossil fuels. With all-time-high levels of carbon emissions from burning fossil fuels precipitating irreversible climate change, we need to examine the past to build upon other ways of enlisting alternative and renewable energy for both work and comfort.

Scientists and climate activists agree that renewable energy sources are critically important to our energy future, but they are hardly new to modern society. Humans have experimented with various renewable energy for decades, if not centuries. Chapter 3 covered how the kinetic energy of moving water can also transform into hydropower electricity through a turbine and generator. Moving air, such as wind, also has kinetic energy that a windmill or wind turbine can harness for work or to generate electricity. The sun does not heat the entire surface of the earth equally, so gases move from high-pressure to low-pressure zones depending upon the topography. Wind is actually the result of solar energy heating the air, causing it to rise, and cold air blowing in to fill the space like a vacuum. Certain areas and topographies produce higher winds than others, often with predictable direction and speed.

The sun provides a direct form of energy that has allowed human beings to develop and thrive across the globe, although its frequency and consistency through the day and seasons varies. The gravity at the sun's core creates so much pressure that it fuses the sun's hydrogen gas into helium. That process creates the light and the heat essential to human life. Both nature and technology can convert solar energy to other forms of energy. Photosynthesis in plants turns solar energy into chemical energy, enabling them to produce their own food. This process uses carbon dioxide from the atmosphere and releases oxygen as a byproduct, and so it is an important tool for countering the carbon created from carbon-based fossil fuels. Solar-powered photovoltaic panels are a technological resource in which photons of sunlight directly excite electrons in silicon cells, using the photons of light from the sun to produce electricity.

Efforts to change how much energy, and which energy sources, we use to fuel our modern life through scientific explanation have yet to persuade much of the public to support a transition from using fossil fuels. As the oceans warm, droughts become longer and more frequent, glaciers melt, and ecologies tilt out of balance, public historians must offer alternative narratives that advocate for and contextualize alternative energy. History museums and sites are only beginning to provide past evidence of energy use and practice. Cities like Alberta, Canada, have provided fairly detailed, historical accounts of energy resources on their websites, but political interests and economic sponsorship often influence interpretation at science museums or energy-themed exhibits.[2] Perhaps enlisting the senses, emotions, and communication skills found in the arts and humanities are more appropriate ways to urge essential changes in use and behavior. This interdisciplinary approach—incorporating art, rhetoric, and comforting narratives of progress and ingenuity—is something that fossil fuel–themed museums have done effectively to not only reflect, but also profoundly influence, energy history and how the public conceives it. Renewable energies need similar champions who can frame energy transition to clean and renewable energy as urgent and inevitable.

Histories and Contexts

As a society, we often fail to consider renewable energy as part of our cultural or technological heritage. To understand the challenges and opportunities associated with energy transition, we must examine science, technology, *and* culture. The sun, provider of warmth and light, acquired supernatural traits for many world cultures. Like water, the wind's power is evidenced across the earth's topography in various landforms that identify places and homelands. Many cultures equate the ephemeral quality of wind with the breath of life, or the presence of spirits or gods. Wind power against sails has allowed humans to navigate bodies of water and migrate across the continents for millennia. Wind-rotated blades create kinetic energy that can operate machinery. People have enlisted wind energy to grind grain into flour, draw water, and even power machines that transformed the economy. Robert Righter observes that engineers of the twentieth and twenty-first centuries have cast aside old sources of energy for the convenience, accessibility, control, and the economy of scale that fossil fuels offer. In his book *Wind Energy in America* (1996), Righter provides a historically contextualized narrative for understanding the development of wind energy's history and technology over the last century. He acknowledges that while a centralized power system based on fossil fuels dominates industrial America, the continued interest in wind energy reflects a population, primarily rural, who rejected the trappings of urbanization, industrialism, and capitalism to remain off the system. The environmental conservation movement in the early twentieth century also reflected those concerned about the impact of industrialization on the environment. Wind offered an alternative to those with a penchant for self-reliance to stay off the main power grid (the electrical distribution infrastructure) but still incorporate modern electrical power into their daily work and domestic lives.

We can look for opportunities about the historic use of wind power through sailboats in maritime museums and through the windmills at historic agricultural sites.[3] Wind is plentiful across the Great Lakes as well as in the arid West, but erosion, tornados, and dust storms plague the region. Issues of reliability and the diversity of designs prevented wind energy from attracting the markets that utility companies needed to sustain profit. Small individual wind power stations that people used to pump up groundwater for domestic or agricultural use were not always reliable. Both private industry and the government (through the Tennessee Valley Authority and the Rural Electrification Administration during the Great Depression) secured our society's adoption of the model of the electrical grid by diminishing technical and financial barriers to energy access and use.

The energy crisis of the 1970s encouraged alternative and renewable energy. Following the incentives of the Public Utility Regulatory Policies Act in 1978, companies developed wind turbines that could feed the grid, not just individuals. But the public's

Figure 7.1. Windmill for pumping groundwater in the Klamath Basin, c. 2000. Oregon Water Science Center, Public Domian, United States Geological Services, https://www.usgs.gov/

embrace of renewable energy remained tepid. Righter laments that much trepidation is cultural. Would-be consumers have complained about both auditory and visual pollution. He explains that "the wind turbines can produce 'landscape guilt' because they violate our historic perception of what landscape ought to be."[4] We still need to reconcile the paradox of the machine in the garden.

Other renewable energies similarly fell out of the general public's energy purview or have remained historically site-specific. Historians have not paid adequate attention to the history of geothermal energy. From Native Americans to ancient Romans to migrant settlers in the American West, people have used the geothermal energy of springs and geysers for centuries for many purposes: health, relaxation, meeting places, and eventually tourism. Places like Yellowstone National Park interpret geothermal energy as a natural phenomenon, where waters are visible and close to the earth's surface. Geothermal energy can tap the water or steam from hot springs just below the earth's surface, or as technology becomes more precise, through drilling.[5] While residents of Chaudes-Aigues, France, created the first geothermal heating district as early as the fourteenth century, twentieth-century communities in Boise, Idaho, and Klamath, Oregon, piped water from the hot springs to community buildings as "district heating systems." In 1904, Italian prince Piero Ginori Conti invented the first geothermal power plant in Tuscany. Iceland also began geothermal district heating in the 1930s, as did The Geysers, an area north of San Francisco, which began as a tourist and health attraction. The geothermal power plants produce a significant portion of California's renewable energy today.[6]

Biofuels have become an increasingly contentious source of energy debate as an example of (similar to nuclear energy) a seemingly "green" renewable fuel that, while offering an alternative to oil, can also have adverse environmental impacts. While monopolies and barons dominated narratives about industrial oil production, narratives about biofuels are associated with the more self-reliant and independent yeoman farmer.[7] Their development has largely occurred within the culture and politics of agrarianism, and, like wind, is specific to certain geographic regions. Author Quentin Skrabec argues that Henry Ford and George Washington Carver tried to create an energy future that favored ethanol and waterpower.[8] Energy historians Jeffrey Manuel and Tom Rogers are currently working on a transnational history of ethanol (across the United States and Brazil). *The Perennial Alternative: A Century of Biofuels in the U.S. and Brazil and Its Lessons for the Future* intends to provide a history for a historic energy source that, similar to wind and solar, we have failed to remember or historicize despite humans' use of it for millennia.[9]

Humankind has enlisted the sun as thermal energy since our beginnings, and John Perlin's *Let It Shine: The 6,000-Year Story of Solar Energy* is as comprehensive as it sounds. Perlin provides a history of solar electric technology (engines to photovoltaic), but also of passive solar heating. In the mid-nineteenth century, the French developed solar motors. In places along the East Coast, Americans converted the sun's heat energy to make salt from brine water in places that included Cape Cod and Syracuse, New York. John Ericsson channeled the heat of the sun using a mirror for a hot-air combustion engine, but not only was use thus restricted to the daylight hours, the engines were just too expensive for marketability. Aubrey Eneas developed a solar motor for pumping up groundwater and found eager customers in the sunny Arizona deserts where surface water was sparse and where groundwater needed heavy pumping for irrigation purposes. Cost was again an issue, as well as the size and fragility of the mechanism. By the turn of the twentieth century, European scientists also began giving solar energy serious thought. Solar water heaters worked in warm regions, but again, electrical versions were far more convenient and inexpensive, and manufacturing became very difficult when the total war efforts of World War II limited copper use in manufacturing to military purposes.[10]

Interpreting Clean Energy: Solar and Wind

More so than other energy histories, renewables can convey a nuanced story about energy choices. Both economic and cultural values guided those choices. However, the history of wind energy can explore the broader political ideal of individualism coming into conflict with the utilities' embrace of capitalism, efficiency, and economies of scale that led to today's electrical grid system (see next chapter). Trade and commercial catalogs corresponding to the period of significance of your site can illustrate what choices individuals may have had in types of wind generators and appliances, for example.

This renewable energy material culture that creates wind and solar is important to examine. People have long objected to energy infrastructure like pipelines, smokestacks, and power lines. Daniel Wuebben's work *Power Lined* (2019) explores the connection between aesthetics and perceptions of energy.[11] We are not accustomed to bearing witness to energy production, nor considering that appropriate energy use must balance economies of scale with regionally based cultural and ecological conditions. Aesthetic sensibility is one of the most significant obstacles to the widespread adoption of wind and solar, particularly when those with political and social power consider wind turbines unsightly robots that disrupt viewsheds. Likewise, solar panels alter rooflines and can obscure architectural details. The tension between renewable energies, cultural heritage, and historic preservation requires reconciliation, primarily over site selection. The debate over Cape Wind, the 2010–2013 wind farm project in the Nantucket Sound, is the most prominent example of when preservation and renewable energies could not reconcile.

The eventually defeated project received a lot of press about wealthy Cape Cod residents objecting to the disruption of their ocean views. However, the windmills additionally threatened to interrupt the viewshed of a traditional cultural property, that of the Wampanoag ("People of the First Light"), whose culture is tied to it. The lawsuit asserted that:

> The Project will harm the Tribe's religious, cultural, and economic interests by degrading the Nantucket Sound ecosystem and, in particular, disturbing the currently unblemished view of the eastern horizon, both of which are of immense spiritual importance to the Tribe; by disrupting or preventing fishing on Horseshoe Shoal (within Nantucket Sound) as a source of sustenance, subsistence, and income for individual tribe members; and by disturbing the sea bed, which may result in irreparable damage to historically significant and culturally and spiritually important archeological resources.[12]

692 SEARS, ROEBUCK & CO., Cheapest Supply House on Earth, Chicago. CATALOGUE No. 111.

USEFUL INFORMATION PERTAINING TO WINDMILLS.

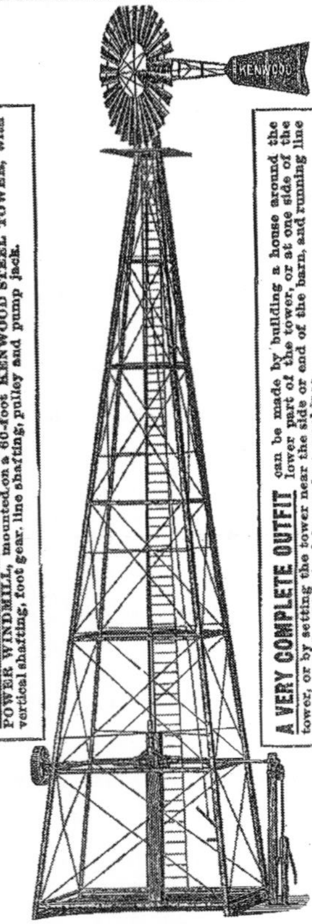

WE SHOW IN THIS ILLUSTRATION a simple, but very complete Power Mill Outfit, consisting of a KENWOOD POWER WINDMILL, mounted on a 60-foot KENWOOD STEEL TOWER, with vertical shafting, foot gear, line shafting, pulley and pump jack.

A VERY COMPLETE OUTFIT can be made by building a house around the lower part of the tower, or at one side of the tower, or by setting the tower near the side or end of the barn, and running line shafting with which to drive various machines.

Diameter of Wheel, Size of Cylinder and Size of Suction Pipe, Which We Recommend for a Given Depth of Well or Elevation of Water.

Diameter of Windmill Wheel	10 Feet Elevation of Water				25 Feet Elevation of Water				50 Feet Elevation of Water			
	Length of Stroke	Diameter of Cylinder	Size of Suction Pipe	Gallons Per Hour	Length of Stroke	Diameter of Cylinder	Size of Suction Pipe	Gallons Per Hour	Length of Stroke	Diameter of Cylinder	Size of Suction Pipe	Gallons Per Hour
Ft.	In.	In.	In.		In.	In.	In.		In.	In.	In.	
8	....	....	....	....	8	3½	1½	800	8	3	1¼	586
8	6	5	3	1224	6	4	2	784	6	3¼	1¼	518
8	....	....	....	....	....	....	....	....	4½	3½	1¼	450
9	....	....	....	....	8	4	2	1045	8	3¾	1¼	690
9	6	6	3½	1763	....	....	....	....	6	3½	1½	600
9	....	....	....	....	4½	5	3	918	4½	4	2	588
10	8	6	3½	2418	....	....	....	....	8	3½	1½	800
10	....	....	....	....	6	5	3	1224	6	4	2	784
10	....	....	....	....	4½	6	3½	1322	....	....	....	....
12	12	8	5	7268	12	5	3	2448	12	4	2	1568
12	....	....	....	....	9	6	3½	2644	....	....	....	....
12	....	....	....	....	....	....	....	....	6	5	3	1224

Diameter of Windmill Wheel	75 Feet Elevation of Water				100 Feet Elevation of Water			
	Length of Stroke	Diameter of Cylinder	Size of Suction Pipe	Gallons Per Hour	Length of Stroke	Diameter of Cylinder	Size of Suction Pipe	Gallons Per Hour
Ft.	In.	In.	In.		In.	In.	In.	
8	8	2½	1¼	408	8	2¼	1¼	330
8	6	2¾	1¼	371	6	2½	1¼	306
8	4½	3	1¼	330	4½	2¾	1¼	278
9	8	2¾	1¼	495	8	2½	1¼	408
9	6	3	1¼	440	6	2¾	1¼	371
9	4½	3¼	1¼	389	4½	3	1¼	330
10	8	3	1¼	586	8	2¾	1¼	495
10	6	3¼	1¼	518	6	3	1¼	440
10	4½	3½	1½	450	4½	3¼	1¼	389
12	12	3½	1	1200	12	3	1¼	880
12	9	4	2	1176	9	3½	1½	900
12	....	....	....	....	6	4	2	784

Diameter of Windmill Wheel	125 Feet Elevation of Water				150 Feet Elevation of Water				200 Feet Elevation of Water			
	Length of Stroke	Diameter of Cylinder	Size of Suction Pipe	Gallons Per Hour	Length of Stroke	Diameter of Cylinder	Size of Suction Pipe	Gallons Per Hour	Length of Stroke	Diameter of Cylinder	Size of Suction Pipe	Gallons Per Hour
Ft.	In.	In.	In.		In.	In.	In.		In.	In.	In.	
8	8	2	1¼	261	....	....	....	....	....	....	....	....
8	6	2¼	1¼	248	6	2	1¼	196	....	....	....	....
8	4½	2½	1¼	230	4½	2¼	1¼	180	4½	2	1¼	147
9	8	2¼	1¼	330	8	2	1¼	261	....	....	....	....
9	6	2½	1¼	306	6	2¼	1¼	248	6	2	1¼	196
9	4½	2¾	1¼	278	4½	2½	1¼	230	4½	2¼	1¼	186
10	8	2½	1¼	408	8	2¼	1¼	330	8	2	1¼	261
10	6	2¾	1¼	371	6	2½	1¼	306	6	2¼	1¼	248
10	4½	3	1¼	330	4½	2¾	1¼	278	4½	2½	1¼	230
12	12	2¾	1¼	741	12	2½	1¼	612	12	2¼	1¼	496
12	9	3	1¼	660	9	2¾	1¼	556	9	2½	1¼	459
12	6	3¼	1¼	518	6	3	1¼	440	6	2¾	1¼	371

Diameter of Windmill Wheel	250 Feet Elevation of Water				300 Feet Elevation of Water			
	Length of Stroke	Diameter of Cylinder	Size of Suction Pipe	Gallons Per Hour	Length of Stroke	Diameter of Cylinder	Size of Suction Pipe	Gallons Per Hour
Ft.	In.	In.	In.		In.	In.	In.	
8	....	....	....	....	....	....	....	....
8	....	....	....	....	....	....	....	....
8	....	....	....	....	....	....	....	....
9	....	....	....	....	....	....	....	....
9	....	....	....	....	....	....	....	....
9	4½	2	1¼	147	....	....	....	....
10	....	....	....	....	....	....	....	....
10	6	2	1¼	19	....	....	....	....
10	4½	2¼	1¼	186	4½	2	1¼	147
12	12	2	1¼	392	....	....	....	....
12	9	2¼	1¼	372	9	2	1¼	294
12	6	2½	1¼	306	6	2¼	1¼	248

In the above table we show the approximate amount of water elevated per hour by different sizes of windmills using different sizes of cylinders and at various strokes. This table is based upon the average amount of water the different windmills will lift with average winds to operate them when windmills and pumps are properly set and connected. In other words, it is estimated that, taken the year around, the windmill will elevate the amount of water shown in table on an average of eight hours per day. Capacity of pumping windmills will vary under different conditions, but whatever variations there may be, we guarantee that the Kenwood Windmill will pump as much water under the same conditions as any windmill of corresponding size and that they will operate in much lighter winds than most other windmills.

The heavy faced figures in the table show the size of cylinder, size of suction pipe and the length of stroke which we recommend for use in connection with the various windmills at stated elevations. For elevating water over 100 feet the work is too heavy and the strain too great for ordinary cylinders and small pipe. Deep well cylinders should be used for greater depth and in such cases whenever possible you should use pipe larger than the cylinder.

For general work do not select a cylinder larger than those recommended for 50-foot elevation. Cylinders larger than 4 inches cannot be used with ordinary pumps. 5-inch, 6-inch and 8-inch cylinders are used for irrigation purposes or in shallow wells only, and require special piping and fittings above the cylinders.

It has been found by experience that it requires, on an average, a wind of a velocity of four to five miles per hour to drive a steel windmill, and that the windmill will run, on an average, eight hours per day. The average velocity of wind in the United States is sixteen miles per hour for eight hours per day. From this it is quite evident that a windmill is a profitable investment, because when a good windmill is properly erected it becomes a faithful and reliable servant, **upon which you can depend for an average of eight hours of steady work for every day in the year, requiring no feed, no fuel, and but very little oil and attention.**

Table of Wind Velocity and Pressure.

Velocity of Wind: Miles per Hour	Velocity of Wind: Feet per Second	Pressure: In Lbs. per Sq. Ft.	Force of Wind	Velocity of Wind: Miles per Hour	Velocity of Wind: Feet per Second	Pressure: In Lbs. per Sq. Ft.	Force of Wind
1	1.47	.005	Hardly perceptible.	30	44.01	4.429	High wind.
2	2.93	.020	Just perceptible.	35	51.34	6.027	
3	4.40	.044		40	58.68	7.873	Very high storm.
4	5.87	.079	Gentle, pleasant wind.	45	66.01	9.963	
5	7.33	.123		50	73.35	12.300	
10	14.67	.492	Pleasant, brisk gale.	60	88.02	17.715	Great storm.
15	22.00	1.107		70			A hurricane that blows down trees, buildings, etc.
20	29.34	1.968	Very brisk.	80	117.36	31.490	
25	36.67	3.075					

Weight of Water in One Foot Length of Pipe.

Size	Lbs.	Size	Lbs.	Size	Lbs.	Size	Lbs.
½-in.	.086	1½-in.	.774	3 -in.	3.087	4½-in.	6.966
1 -in.	.343	2 -in.	1.372	3½-in.	4.214	5 -in.	8.575
1¼-in.	.537	2½-in.	2.159	4 -in.	5.488	6 -in.	12.348

Amount of Water Discharged by One Inch of Pump Stroke.

Diam. cylinder, inches..	2	2¼	2½	2¾	3	3¼	3½	4	5	6	8
Gallons, per stroke......	.0136	.0172	.0212	.0257	.0306	.0359	.0416	.0544	.0850	.1224	.2176

To ascertain the capacity of a pump with any diameter of cylinder given in above table, multiply the gallons per stroke, as given in the table, by the length of pump stroke, and the result thus obtained by the number of pump strokes per minute. This result multiplied by sixty will give the capacity in gallons per hour.

The capacity of single and double acting pumps is precisely the same, the difference being that the single acting pump discharges all the water on the up stroke of the pump rod, while the double acting pump discharges half on the up stroke and half on the down stroke of the pump rod.

IN THIS ILLUSTRATION

we show the Kenwood Power Windmill as it appears when mounted on a wood mast over the barn, showing the manner of bracing the mast with our steel guy rods and how various machines may be driven by the power developed by the windmill.

FOR PRICES SEE PAGES 693 TO 695.

Figure 7.2. Catalog image steel windmill. Product description includes serval pages of details on construction. *The 1902 Edition of the Sears Roebuck Catalog* (New York: Bounty Books, 1969)

In 2016, an appeals court eventually ruled that the Bureau of Ocean Energy Management had not obtained "sufficient site-specific data on seafloor" as obligated by the National Environmental Policy Act.[13]

The multifaceted objections to projects like Cape Wind should serve as contexts for managing historic properties and mitigating the impact of renewable energy projects. More so, its example shows that engaging in early outreach to stakeholders, discussions, negotiations, and compromises can achieve renewable energy goals and environmental justice by honoring those historic properties and landscapes. Back in 2014, Steve Burg wrote a blog about similar conflicts in Pennsylvania and Sweden.[14]

CASE STUDY: Siting Wind Farms; From Steven Burg, "Get Your Wind Farm Off My Historic Site: When Visions of Sustainability Collide"

Off the east coast of Southern Sweden, a battle is raging between competing visions of sustainability. On the most unlikely of battlegrounds, bucolic Öland Island, a desire to promote renewable energy has brought local officials committed to promoting a sustainable society into conflict with island residents, preservationists, farmers, environmentalists, and local business owners who believe that protecting the island's character and cultural resources is incompatible with a proposal to expand industrially generated wind power on the island.

Öland is a rugged and beautiful strip of land surrounded by the Baltic Sea and linked to the mainland by the six-kilometer-long Öland bridge. The island is dominated by a limestone plateau, sandy beaches, and a three-kilometer-wide swath of arable land along the southwest shore called *Västra Landborgen*. This place has supported human settlements for over five thousand years. It is now a popular summer tourist destination and a favorite getaway of the Swedish royal family who maintain a palace, *Soliden*, on the island.

Due to its relative isolation and limited development, portions of the island retain much of their medieval historical character. Today, the island's farming communities continue to reflect land use patterns established approximately a thousand years ago, including linear villages and a continued division of agricultural lands into "infields" and "outfields." The limestone plateau, *Stora Alvaret*, remains largely undeveloped and home to a unique and diverse island ecosystem.

In 2000, UNESCO's (United Nations Educational, Scientific, and Cultural Organization) World Heritage Committee recognized the island's transcendent value as a cultural resource by inscribing the "Agricultural Landscape of Southern Öland Island" on its list of World Heritage Sites. Öland is also special in another way: Its strong winds and the flat, open landscape of its limestone plateau make it an ideal site for generating wind power. Wind power has been actively developed in Sweden, particularly since the nation adopted a new national energy policy in 2009 that focused on achieving "ecological sustainability, competitiveness and security of energy supply." In 2012, Kalmar County officials adopted a bold Island Sustainable Energy Plan that sought a 50 percent reduction in carbon emission by 2020.

To reach those goals, local officials are planning new industrial wind farms on the island—including one that will position a pair of 150-meter-tall wind turbines within the boundaries of the World Heritage site. In response to concerns raised by UNESCO about the impact of the project, the county governor of Kalmar convened a meeting with "local and regional decision makers" as well as the local World Heritage site's Management Council. At that meeting, all parties invited to the meeting concluded that the project would have no impact on the World Heritage property.

However, some local residents disagreed, and a group of 112 Öland residents petitioned UNESCO, protesting that the project posed a serious threat to the World Heritage site. In a letter written to UNESCO World Heritage Centre director Kishore Rao, the group noted that UNESCO had recognized Öland as a site where "land use has changed little since the Stone Age," a standard they felt should be maintained. They also felt that Sweden had ignored its treaty obligations under the World Heritage Convention of 1985 that required signatories to take legislative and administrative measures to protect cultural and natural resources.

The petitioners also raised concerns about the impact the wind farm might have on heritage tourism. They cited a survey that showed that the World Heritage Site had surpassed Öland's beaches as the most attractive draw for tourists. As Fritz Eriksson, the first signatory of the letter to UNESCO, explained to a reporter from Swedish Radio, "People come for the untouched nature and the stunning scenery. So reasonably, Öland will be worse off if the tourists do not come."

After evaluating the wind farm project, UNESCO officials concluded that placing the turbines within Öland's World Heritage site would "destroy the actual reason that justified the inclusion into the list." If the project were completed, UNESCO warned, "The Agricultural Landscape of Southern Öland will be at high risk to lose the grounds of its nomination and Outstanding Universal Value." Nevertheless, the project continues to move forward.

The case of Southern Öland provides a rather dramatic case where visions of heritage preservation and renewable energy development collided, but it is certainly not unique. Other communities have faced similar challenges, including the World Heritage sites of Mont-Saint-Michel in France (where an offshore wind project was blocked by the French courts), and Britain's Jurassic Coast. In the United States, the Cape Wind project proposed for Nantucket Bay off Martha's Vineyard has generated a fierce legal and political struggle that has spanned more than a dozen years. Cape Wind's Construction and Operation plans received approval from the U.S. Department of the Interior's Bureau of Ocean Energy Management, Regulation and Enforcement in 2011 despite a determination by the Advisory Council on Historic Preservation that the wind farm would negatively affect thirty-six historic sites and districts, and six resources of cultural and religious significance to the region's Indian tribes.

The many compelling arguments in favor of renewable energy projects makes opposing them challenging. Renewable energy is widely viewed as an environmentally friendly source of power that enables the comfortable standard of living to which modern societies have grown accustomed and supports energy independence without the toxicity of fossil fuels. Wind turbines and solar panels also allow nations to continue generating energy while reducing carbon emissions, something particularly important for nations such as Sweden that are signatories to the 2005 Kyoto Protocol, and that also must comply with the European Union's nation-specific targets for reducing carbon emissions set in 2007 (targets that are now being reconsidered as a result of the economic downturn). Commentators such as Thomas Friedman have also argued that a failure to embrace clean and renewable energy strengthens authoritarian, oil-rich regimes.

When such policies are embraced by powerful political leaders and multinational energy corporations, the rhetorical, financial, and political power behind such initiatives can be considerable. Advocating for the protection of cultural resources when the alternative is framed as ensuring national security, protecting the environment, and safeguarding the future of the planet is certainly a daunting challenge.

So how should historic preservationists respond to these threats to cultural resources, particularly when preservationists themselves may favor the development of renewable energy resources? Is there a way to reconcile the competing demands of cultural heritage preservation and renewable energy development?

Yes, but it will require even greater levels of political engagement, particularly at the earliest stages of project development. As Richard Wagner, director of Goucher College's historic preservation program, has noted, preservationists need to gain a seat at the table. They need to play a role when sustainable energy plans and legislation are drafted at all levels of government. In the case of wind and solar projects, they need to advocate for cultural and historical resources when developers are considering sites for prospective projects. They need to work with environmental organizations and other stakeholders to educate local planning offices, elected officials, energy companies, and developers about the value of cultural and historic resources, and to build an ongoing consensus that projects should be developed at a scale, and in locations, that are compatible with protecting cultural and natural resources. Everyone at the table needs to believe that there are some sites that are simply too precious to compromise, even in pursuit of renewable energy.

Historic preservation advocates also need to muster strong arguments in defense of the preservation ethic. These arguments need to be based on aesthetics and economic arguments, as well as the value of preserving communities' sense of place. However, historic preservation advocates need to clearly articulate the value of historic preservation in building a sustainable society. For historic structures, such arguments may focus on the legacy investment of money, energy, and natural resources in existing structures; the potential for historic structures to provide affordable housing; or the way that restoration projects can create jobs for local businesses and craftsmen. Historic landscapes, especially those of pre-industrial agricultural societies, must also be viewed as precious cultural resources that can be studied to reveal lessons about sustainable land use techniques, construction methods, technologies, and patterns of social organization that might be worth emulating or adapting to contemporary societies.

Historic preservationists also need to demonstrate that all renewable energy projects are not created equal. Some can be constructed with minimal impact on a region's environment or quality of life. However, large-scale industrial wind turbine projects require significant amounts of clearing, blasting, and leveling to create turbine pads, access roads, and power lines. They also have the potential to create noise and traffic that can dramatically alter a region's ecosystem and soundscape. Such projects can often be driven less by concern for the environment than by the promise of profits to be reaped by large-scale projects underwritten with generous public subsidies. The rush by energy corporations to capitalize on government economic incentives without full consideration of the impact that renewable energy projects can have on communities, cultural, and natural resources has led groups in North America and Europe to aggressively oppose wind farms and to even pursue local moratoriums on wind development.

In the absence of good policy and process, preservationists and communities must identify the sites of greatest value for protection and work to mitigate the impact of renewable energy development at other locations. When projects are underway that threaten significant cultural resources, preservation coalitions must decide if they will take steps to stop a project. Such fights are hard, but on occasion, they can be won.

One example is the case of Dutch Corner in rural Bedford County, Pennsylvania. A large wind farm project slated for Evitts Mountain threatened to undermine a historic landscape encompassing more than thirty historic farms and related buildings, leading to the site's listing on Preservation Pennsylvania's 2010 "Pennsylvania at Risk" list. By building a coalition of local and national organizations, creating a National Register Historic District, winning passage of a strong local preservation ordinance, and blocking efforts to weaken the ordinance and shrink the historic district's boundaries, community members mounted a vigorous campaign to stop the project.

Yet the factor that ultimately stopped the wind farm, according to Iberdrola Renewables spokesman Paul Copleman quoted in the *Pittsburgh Post-Gazette*, was the end of federal tax credits for renewable energy projects at the end of 2013. The site was saved, but it is hard to know what the outcome might have been if federal tax policy had not changed.

Ultimately, the best way to advance a vision of a sustainable society that values clean energy development, and the protection of cultural resources, is to establish public policies at the local, state, national, and international levels that recognize the value of both and establish frameworks for how they might best be balanced in decision-making. This is a matter of public education but, more significantly, a matter of political advocacy. Political leaders responsible for policymaking need to recognize that heritage preservation does not need to be viewed only as an obstacle to energy development initiatives but rather as a complementary and essential component of a sustainable society.

CASE STUDY: Connecticut Trust for Historic Preservation (now Preservation Connecticut), Hamden, Connecticut

We can practice historic preservation and interpretation in harmony with energy technology. In 2016, the main office of the Connecticut Trust for Historic Preservation installed solar photovoltaic panels on its main offices, the historic boarding house of the Eli Whitney Armory historic site. The panels serve as an example of technological innovation, which is also the primary theme of the Eli Whitney Museum and Workshop, of which the boardinghouse is a contributing resource (see chapter 3). While not currently interpreted as part of the historic site across the street, a partnership between the preservation organization and the historic site could collaboratively interpret the history of energy at

Figure 7.3. Solar Panels on Whitney Boarding House, Hamden, Connecticut. Preservation Connecticut, Hamden, CT

the site together. The Trust attested that, "By using the boarding house as a demonstration project for the use of sustainable technologies on historic buildings, the Trust is carrying on Whitney's legacy of innovation." Updating physical resources to meet current needs can tell multiple and interdisciplinary tales that can help inform future decision-making and offer educational opportunities for the public. As the Regional Water Authority prepares to rehabilitate the historic dam at the Eli Whitney Historic Site, the task offers interpretive opportunities to talk about water use and energy diversity, both historically and in the present.

The solar panels reduce reliance on carbon-based power sources, while "publicly demonstrating the compatibility of historic structures and sustainable technologies."[15] When attached to homes, solar panels are removable and thus are not irreversible alterations, so most preservation councils and historic preservation offices consider them fairly harmless. The National Park Service has published substantial guidance on the matter. This is especially true if one cannot see the panels from a public right-of-way, they blend with the color of the roof itself, and they are arranged in a way that does not disrupt the historic, character-defining features, particularly the roofline.[16]

Artifact Spotlight

Windmill Pump and Wind Turbine

Tracing the material culture of windmills can actually interpret both wind and electricity, as well as highlight the significance and influence of energy on Western expansion. Industrialists and individuals had used steam-, horse-, and waterpower to mechanically operate stationary machinery. In New England, plentiful rivers made waterpower a favored energy source for milling, so it is somewhat noteworthy that Daniel Halladay, a machinist from Connecticut, developed a windmill in 1854 with a "centrifugal governor" to regulate the speed of the blades. The device could also stop when winds became too high, preventing damage.[17] Halladay's advertisement boasted, "Steam, horse, and water-power, have been variously used for driving stationary machinery. The two former require the expenditure of fuel or feed, and the latter does not exist on many farms, and can only be occasionally used. But there is another—and universal power—found on every part of every single farm in the world—and sweeping over all with the strength of a thousand horses—which has been very little used for farm purposes. This is the wind."[18]

The U.S. Wind Engine and Pump Company purchased the Connecticut-based Halladay Windmill Company and moved it to Batavia, Illinois, a popular center of windmill production, in the Fox River Valley. Just thirty-five miles west of Chicago, the Fox River Valley had abundant limestone for building factories and waterpower from the river.[19] Due to the initial cost of steel manufacturing, windmills made from iron and steel did not become commercially profitable until the turn of the century. The U.S. Wind Engine and Pump Company marketed its galvanized steel towers through its own descriptive catalogs, dealers, agents, or in exhibitions to an expanding market of Western settlers needing to pump water for farming in the arid region.[20] Hence, the windmill design of Connecticut encouraged Western expansion by enabling energy from wind to more efficiently pump groundwater to support homesteads and commercial agriculture, as well as carrying the hydropower of rivers from reservoirs to cities and towns to electrify the region.

In the late nineteenth century, the design of windmills became increasingly efficient. They allowed rotational energy to generate electricity, first in Glasgow, Scotland. Poul la Cour, a Danish scientist further refined the designs. Jacobs Wind Electric Company, which began in the 1920s to sell wind power plants to individual farms and ranches across the upper Midwest, provides a useful web page reviewing the history of its own company with a source bibliography.[21] Such designs, however, could not sustain the investment necessary to perfect them, especially in competition with the consistency of seemingly (but falsely), inexhaustible fossil fuel systems.

Interestingly, places with the highest fossil fuel consumption also lead in wind energy production. Famed for oil, Texas also leads the other states in wind energy production, with its one major museum on windmill history dwarfed by the thirty or so museums featuring petroleum exhibits. The American Windmill Museum (aka American Wind Power Center) in Lubbock, Texas, displays water and electrical windmills. The museum, established in 1997 by a Texas Tech instructor and an industry CEO, illustrates the evolution of wind technology from small-scale irrigation to turbines.[22]

Beginning in 1994, the city of Batavia, Illinois, began to locate, purchase, renovate, and erect several original locally manufactured windmills to display along the Batavia Riverwalk, an integral part of the heritage landscape and identity of the city. Signage accompanies each of the seventeen structures. A publication produced by the American Society of Mechanical Engineers provides a basic technological overview.[23] In 2021, the York County History Center in York, Pennsylvania, launched a digital *StoryMap* exhibit of the Smith-Putnam Wind Turbine, whose eleven-by-sixty-five-foot blades spun from 1941 to 1945. The city is home to the S. Morgan Smith Company, known for its hydro turbines, and the History Center holds the company's records.

Inspired by designs in Russia on the eve of America's entry into World War II, the Smith-Putnam Wind Turbine aimed to produce enough power to serve a whole town. Such a timely historical subject digitally presents intriguing images in a virtual exhibit to highlight this important effort at generating electricity from wind on a large scale. However, the exhibit lacks broader context and themes beyond business and technology, likely because it is based on the company archives. The Smith-Putnam Wind Turbine Collection has many examples of advertisements and catalogs that sold at-home wind turbines. These resources allow one to compare different designs, but also consider who companies may have marketed to, and again, consider the user as much as the maker. The Smith-Putnam Wind Turbine showed that large-scale wind power generation was possible. However, by the time it failed from a cracked blade and financial insolvency, loss of confidence and the promise of nuclear and other power

Figure 7.4. The world's first megawatt-size wind turbine on Grandpa's Knob, Castleton, Vermont, 1941. Public Domain, PD-US-GOV-DOE, www.nrel.gov

sources overshadowed any further private and government investment. A discussion of the hydroelectric turbines for which the company was known could offer a comparison that informs visitors more about energy diversity than focusing on a singular technology does. Enlisting secondary sources to provide explanations for why this innovative and effective invention failed to develop would further sustain attention.[24]

T. Lindsey Baker's *A Field Guide to American Windmills* (1985) and Richard Leslie Hill's *Power from Wind* (1996) provide good resources for reviewing these artifacts.[25] Toy models of windmills, wind turbines, and solar-powered toys can also provide hands-on demonstrations about these technologies, but it is important to show images of historical precedents alongside them. While some of the specific technologies are newer, enlisting wind and solar energy for work is hardly new, nor is controlling these energies for domestic use.

Historic Homes

Many historic houses illustrate passive wind, solar, and in-ground energy systems through their very design, with vernacular architectural styles reflecting the climate of the region, rather than reflecting cultural or artistic trends as we see in the twentieth century. Homes enlisting passive solar energy date back to ancient Chinese, Egyptian, and Greek architectural styles that reflected their regional climates. Windcatchers, for example, are tower- or chimney-like architectural features common in West Asia and Africa that capture cool breezes through openings located opposite that of the direction that the wind usually blows, and they funnel the cool air down through a shaft. Warmer air also rises through these features out of the living space.[26]

In the desert of the American Southwest, thick adobe walls kept interiors relatively cool, and groups like the ancient Puebloans situated their dwellings to maximize the heat of the winter sun. The door of the traditional hogan of the Diné (Navajo) faces the sunrise, a central stove serves as a heating source, and a hole in the roof provides ventilation. Mud offers insulation and climate mediation for the cedar and pine log frame. Temporary summer dwellings served to shade inhabitants, but some also adopted cupolas with louver-like shutters that worked like windcatchers. Newer housing has adapted physical forms to the needs of electricity. Those unable to connect to the electrical grid (see next chapter) of transmission and distribution lines have begun to install solar panels, allowing for the more traditional dispersed settlement patterns rather than the clustering required by central station power. However, large numbers of off-grid residents still lack such modern amenities.[27]

There are many interesting places for historic sites to incorporate a discussion of wind and solar energy through architectural features, breathable and flexible materials, and designs that regulated heating and cooling. Industries, developers, realtors, and the general home-owning public consider older homes to be energy inefficient. This is because often the people living in these places today want older houses to conform to newer heating and cooling technology for which they were never designed. Rather, architects designed houses as holistic energy systems in and of themselves. In the eighteenth century, "builders sited houses to capture winter sun and prevailing summer cross breezes," and they arranged windows, doors, and the walls to "maximize ventilation."[28] Popular New England saltbox designs often faced toward the sun with windowed two-story front rooms, and the long sloping roof on the back providing protection from winds. A vined trellis on the exterior could shade the rooms in the summer. The central hall layout of a traditional Colonial-style home promotes whole house heating through the centrally located fireplace. The thick walls and natural insulation of sod houses of prairie settlers in the nineteenth century similarly provided cooler interiors in the summer. These shelters also captured heat from a stove fueled by wood, or more likely hay, straw, or dried animal dung (aka cow "patties" or "chips") in the winter. Subterranean "root" cellars provided refrigeration and necessary humidity to preserve foods like vegetables, fruits, nuts, and even animal feed. These could be sited near or away from the house. The ice industry delivered its product daily in the nineteenth and early twentieth century to provide cooling before refrigerants, which are significant greenhouse gases.

Even in the nineteenth century and in urban areas where lot size limited choices in building orientation, architectural elements aided in forced air ventilation. Windows allow the entrance and retention of solar heat when closed, and they draw in cooling wind when opened. Especially in Victorian era Queen Anne and early-twentieth-century craftsman homes, architects strategically arranged large windows to encourage the breeze and mimic a

Figure 7.5. "Art Work of Toledo, Ohio," featuring sleeping porch on Victorian home, c. 1895. Published by W. H. Parrish Publishing Company (Chicago), Donation from collection of Toledo-Lucas County Public Library, to Digital Public Library of America.

wind tunnel. Other examples of this were the shotgun house in the Deep South and houses that featured an open space through the middle, known as a "dogtrot," found in regions with warm climates. High ceilings, louvers, transom windows, open floor plans, or even a window at the top of a wide-open stairwell, for example, will draw up hotter air and funnel it out an upstairs window, resulting in a breeze. Covered porches kept the air just outside of the house cooler so windows could remain open even when raining.[29]

Extended shelters, overhangs, awning, and even drapes could provide shade, shutters, outdoor living space, and cool the air coming directly into the house. Those same overhangs could trap the lower winter sun indoors as "reradiated" heat. Skylights and cupolas invited the winter sun inside and kept interior spaces protected in the summer.[30] Unlike the front porch, which encouraged socialization, the screened or sleeping porch in the Southern and Western regions of the United States maintained breezy conditions. The National Association for the Study and Prevention for Tuberculosis published a book in 1912 entitled *Fresh Air and How to Use It* in which they expound on the many benefits of sleeping porches, tents, and open-air bungalows. Screened in and located to receive breezes in all directions, a sleeping porch could augment the airflow, helping control airborne diseases like cholera and tuberculosis before advances in vaccines and antibiotics.[31] President Taft even built one for the White House.[32]

In addition to Perlin's chapter on solar house heating, Daniel Barber's *A House in the Sun: Modern Architecture and Solar Energy in the Cold War* examines the mid-century modern architecture specifically designed to capture solar energy. He not only explores the design, but also the marketing in popular magazines like *Life* and *Ladies' Home Journal*. Barber credits the impetus for these designs to the Cold War. Energy security concerns encouraged experimentation with certain features, including southern exposure façades, flat roofs, and open floor plans. Eventually, the mechanical systems based on central furnaces offered more flexibility in home design as priorities shifted to accommodate cultural values and lifestyles through home design.[33] In an online tour of Cleveland, Ohio, Richard Raponi provides historical perspective on images of a model home built for the Home and Flower Exposition in 1976 to illustrate energy-saving features, including renewable energy. Although this house became a nature education center when the energy crisis abated, there are no doubt several examples of similar structures that provide ample interpretive opportunities to discuss the promise of renewables.[34]

Original home energy systems become compromised when we replace parts of them piecemeal. Historic sites should model the rehabilitation and renovation of historic properties. For example, original windows are customized to fit the original framing. With glazing, weatherproofing can address places of air infiltration such as sashes and meeting rails. When a modern vinyl replacement window bows as a result of moisture in a settling wood-frame house, it creates a gap around the framing. The insulated glass of modern windows additionally limits visible light and views by requiring thicker mullions and framing.[35]

Historically, plenty of models for renewable energy make it clear that these are not new and experimental technologies, but they are part of our heritage. Renewable energies have history, but they are also important to our energy future. Energy diversity may be as important as energy transition, but each renewable fuel source has its proponents and detractors. Consumer objections to these technologies are primarily aesthetic and auditory, and they are arguably tied to their visibility on the landscape as in the case of solar photovoltaic panels and wind turbines. Still, resistance to alternative energy goes well beyond issues of self-centered "NIMBYism," (Not in My Backyard). Concern for bird safety also drives opposition to wind, and environmentalists argue that the very process of producing biofuels contributes to greenhouse gas emissions, causing ecological damage in soil erosion, fertilizer runoff, and a diminishing affordable food supply. Some object to the potentially toxic chemicals in the materials of solar panels. Large-scale fields of solar panels also threaten wildlife. In many spots of the forested Northeast, some communities are paradoxically proposing clearing wooded areas, where trees absorb carbon dioxide, for space for solar fields.[36]

Finally, there are certainly technical barriers to abandoning fossil fuels for clean energy, including control, storage, and hence reliability. Unlike water, we cannot stop the wind from blowing. Gusts are inconsistent in force and frequency, and one cannot own, contain, or regulate wind's use in the same way that one can with water, land, oil fields, or mines.[37] Batteries can store solar or wind energy as chemical energy, but these technologies still require further development. For these reasons, we need to focus future efforts on developing new technologies and infrastructure that make economic sense, but are also reliable, environmentally sustainable, and equitable.[38]

As we lurch into a necessary future where we must transition from fossil fuel dependence, it is critical to understand why renewables, namely wind and solar, have repeatedly failed to become a significant portion of our energy supply. Fossil fuels and the efficient systems we have developed for extracting and applying them has encouraged energy demands and created a high energy–dependent society and economy. Furthermore, although wind and solar might be viable sources to replace coal for electrical use, shifting away from gas-powered cars will require a titanic and coordinated effort to develop adequate infrastructure. Renewable energies cannot yet meet vast shipping needs over land, water, or air. We did not design our technological and economic systems for them.[39]

The next chapter will review how investment in and design of infrastructure has limited consumer energy choices to primarily fossil fuels and become a barrier to energy transition. Involving engineers, industry, and ecologists, as well as users, in designing efficient, renewable energy infrastructure can help avoid the obstacles and inequities associated with past systems of energy distribution. An informed and motivated public can enlist our considerable consumer power to participate in and advocate for decisions to invest in efficient, safe, and accessible renewable energy infrastructure.[40] We need to examine the energy infrastructure we rely upon today to understand what choices we might create in the future.

Notes

1. See later chapter on the discussion of smoke. Also see Robert Righter, *Wind Energy in America: A History*, 34–35.
2. "Alberta's Energy Resources Heritage," accessed January 18, 2022, http://www.history.alberta.ca/energyheritage/sands/default.aspx.
3. Righter, 22.
4. Righter, 60–61, 249, 265, 275.

5. In his 1883 guide, Henry Wisner famously described a setting where one could fish trout out of Yellowstone Lake and immediately cook it in nearby boiling springs "without taking them off the hook." Henry Wisner, *The Yellowstone National Park* (New York: GP Putnam, 1883), 21.

6. Office of Energy Efficiency and Renewable Energy, "A History of Geothermal Energy in America," Department of Energy, Washington, DC, accessed January 19, 2022, https://www.energy.gov/eere/geothermal/history-geothermal-energy-america.

7. Jeffrey T. Manuel, "Lessons from a Forgotten Fuel: Assessing the Long History of Alcohol Fuel Advocacy and Use in the United States," draft manuscript, 2021.

8. Quentin R. Skrabec, *The Green Vision of Henry Ford and George Washington Carver* (Jefferson, NC: McFarland, 2013).

9. Jeffrey Manuel, Southern Illinois University-Edwardsville, accessed January 10. 2022, https://www.siue.edu/artsandsciences/historical-studies/faculty/manuel.shtml.

10. Leah S. Glaser, *Electrifying the Rural American West: Stories of Power, People, and Place* (Lincoln: University of Nebraska Press, 2009), 49–52; Perlin *Let it Shine: The 6,000 Year Story of Solar Energy* (Novato, CA: New World Library, 2013). A teenager in France created the first photovoltaic cell; Nikola Tesla was fascinated with aspects of what he called "radiant energy" in the United States; and Albert Einstein explored the photoelectric effect and won the Nobel prize in 1905. Perlin, 308–14; William B. Meyer, "Why did Syracuse Manufacture Solar Salt?" *New York History* 86:2 (Spring 2005), 195–209.

11. Daniel L. Wuebben, *Power Lined: Electricity, Landscape, and the American Mind* (Lincoln: University of Nebraska Press, 2019).

12. Gale Corey Toensing, "Aquinnah Wampanoag Sues Feds Over Cape Wind," *Indian Country Today*, July 14, 2011.

13. Public Employees for Environmental Responsibility et al., Appellants Town of Barnstable, Massachusetts et al., *Appellees v. Abigail Ross Hopper, acting director, U.S. Bureau of Ocean Energy Management et al.*, Appellees, United States Court of Appeals, District of Columbia Circuit. No. 14-5301, July 05, 2016.

14. Steven Burg, "Get Your Wind Farm Off My Historic Sites," *History@Work*, Parts 1 and 2, Blog, February 10, 2014. Accessed January 2, 2022. https://ncph.org/history-at-work/wind-farm-historic-site-part-1; see Allison M. Dussias, "Room for a (Sacred) View? American Indian Tribes Confront Visual Desecration Caused by Wind Energy Projects," *American Indian Law Review* 38:2, 2014.

15. "Solar Panels on Historic Properties," Technical Preservation Services, National Park Service, U.S. Department of the Interior, accessed January 10, 2022, https://www.nps.gov/tps/sustainability/new-technology/solar-on-historic.htm. This flexibility toward solar may change as companies like Tesla develop solar shingles; "Solar Energy Comes to the Boarding House," *Connecticut Preservation News* XL:1 (January/ February 2017), 1, 4.

16. "Solar Panels on Historic Properties."

17. Gregg Mangan, "Halladay's Revolutionary Windmill: Today in History," *ConnecticutHistory.org* (August 29, 2020), https://connecticuthistory.org/halladays-revolutionary-windmill-today-in-history-august-29; "Electricity and Alternative Energy: The Halladay and Jacobs Windmills," accessed January 10, 2022, http://www.history.alberta.ca/energyheritage/energy/wind-power/wind-power-in-north-america-and-the-development-of-windpumps/the-halladay-and-jacobs-windmills.aspx.

18. *Illustrated Annual Register of Rural Affairs: A Practical and Copiously Illustrated Register of Rural Economy and Rural Taste* (1858), 222.

19. "Halladay's Revolutionary Windmill."

20. T. Lindsay Baker, *A Field Guide to Windmills* (Norman: University of Oklahoma Press, 1985), 7–10, 22, 70; Leah S. Glaser, "The EMA Transmission Line," (No. AZ-6- B. Historic American Engineering Record, National Park Service, Western Region, 1996), 7–9.

21. Righter, 126–45.

22. Kate Galbraith, "Museum Shows History and Power of Wind Energy," *New York Times*, July 30, 2011, https://www.nytimes.com/2011/07/31/us/31ttwindmill.html; https://windmill.com; other sites include the Wind Energy Museum in the UK, "Wind Energy Museum," accessed January 10, 2022, https://windenergymuseum.co.uk.

23. "Public Art in Chicago," accessed January 15, 2022, http://www.publicartinchicago.com/2017-day-trip-batavia-illinois-the-windmill-heritage-of-batavia; Batavia Historical Society, "Windmills Along the Riverwalk," http://www.bataviahistoricalsociety.org/exhibits-collections/industries-overview/windmills; ASME, "Batavia, Illinois Windmill Collection: A Mechanical Engineering Landmark," brochure, 2013, https://www.asme.org/wwwasme org/media/resourcefiles/aboutasme/who%20we%20are/engineering%20history/landmarks/254-batavia-windmills.pdf.

24. York County History Center, "Energy Awaits: The Smith-Putnam Wind Turbine and the Beginning of Modern Wind Energy in America," accessed February 10, 2022, https://story maps.arcgis.com/stories/7fc2447c836047059b73b002a613d702.

25. T. Lindsay Baker, *A Field Guide to American Windmills* (Norman: University of Oklahoma Press, 1985); Richard Leslie Hill, *Power from Wind: A History of Windmill Technology* (Cambridge: Cambridge University Press, 1996).

26. "Root Cellar," Pennsylvania Agricultural History Project, Pennsylvania Historical and Museum Commission, accessed January 13, 2022, http://www.phmc.state.pa.us/portal/com munities/agriculture/field-guide/root-cellar.html.

27. Stephen C. Jett and Virginia E. Spencer, *Navajo Architecture: Forms, History, Distributions* (Tucson: University of Arizona Press, 1981); Perlin, *Let It Shine*, 3–80; Debra Utacia Krol, "Have You Heard of a Navajo Hogan?" *Architectural Digest* (May 2, 2018), accessed January 31, 2022, https://www.architecturaldigest.com/story/navajo-hogan-eco-friendly.

28. Sharon C. Park, "Heating, Ventilating, and Cooling Historic Buildings," Preservation Brief, accessed January 10, 2022, https://www.nps.gov/tps/how-to-preserve/briefs/24-heat-vent-cool.htm; John A. Burns, *Energy Conserving Features Inherent in Older Homes* (Washington, DC: US Department of Housing and Urban Development and the US Department of the Interior, 1982).

29. Park; Gwendolyn Wright, *Building the Dream: A Social History of Housing in America* (Cambridge, MA: MIT Press, 1981).

30. Perlin, *Let It Shine*, 253–77; Park; Michael Dolan, *The American Porch: An Informal History of an Informal Place* (Guilford, CT: Lyons Press, 2002).

31. Thomas Spees Carrington, "Fresh Air and How to Use It" (Harvard University, 1912), 81–98, https://curiosity.lib.harvard.edu/contagion/catalog/36-990060321790203941; Chris McDaniel, "Hot Summer Nights," *Southwest Living*, accessed January 10, 2022, http://south westlivingyuma.yumawebteam.com/historic/hot-summer-nights.

32. Library of Congress, "Sleeping porch on the roof of the White House erected during the Taft Administration," Washington, D.C., Photograph, 1909, accessed January 3 2022, https://www .loc.gov/item/2002713052.

33. See John Perlin, *Let It Shine*, 219–302; Daniel A. Barber, *A House in the Sun: Modern Architecture and Solar Energy in the Cold War* (New York: Oxford University Press, 2016).

34. Richard Raponi, "Solar Interpretation Center: A Model of Efficiency," *Cleveland Historical*, accessed September 29, 1021, https://clevelandhistorical.org/items/show/700.

35. Walter Sedovic and Jill H. Gotthelf, "What Replacement Windows Can't Replace: The Real Cost of Removing Historic Windows," *APT Bulletin: Journal of Preservation Technology* 36:4, 2005.

36. U.S. Energy Information Administration, "Solar Explained: Solar Energy and the Environment," https://www.eia.gov/energyexplained/solar/solar-energy-and-the-environment.php; Jim Carlton, "Solar Power's Land Grab Hits a Snag: Environmentalists," *Wall Street Journal* (June 4, 2021); Scott Dance, "Go solar, or save the trees? Georgetown University solar farm would clear 240-acre forest in Charles County," *Baltimore Sun* (January 31, 2019); Elizabeth Langhorne and Diane Hoffman, 'No net loss!' Don't cut down the forests to build solar sites," *CT Mirror* (December 1, 2020), https://ctmirror.org/category/ct-viewpoints/no-net-loss -inappropriate-solar-sites-and-forests.

37. Righter, 3–5.

38. National Renewable Energy Laboratory, "Next-Gen Concentrating Solar Power Research Heats Up at NREL: Molten Salts Can Melt Down the Price of Concentrating Solar Power-Plus-Storage," February 9, 2022, NREL: Transforming Energy, https://www.nrel.gov/news/ program/2022/next-gen-concentrating-solar-power-research-heats-up-at-nrel.html.

39. Michael Webber, 238.

40. Christopher Jones, *Routes of Power: Energy and Modern America* (Cambridge, MA: Harvard University Press, 2016); Christopher Jones, "New Tech Benefits the Elite Only Until the People Demand More," *Aeon* (September 6, 2016).

CHAPTER 8

Energy, Access, and Equity

The Infrastructure of Electricity

A FEW MINUTES AFTER four o'clock in the afternoon on August 14, 2003, the lights went out across eight states and two Canadian cities. Without electricity, computers shut down, trains and subways stopped moving, gas pumps stopped pumping, and restaurants purged foodstuffs for lack of refrigeration. As regular daily activities ground to a halt, few impacted would doubt the importance of electricity in the twenty-first century. The Northeast "blackout of 2003" crippled the economy and disrupted the lives of millions. When a power line in Ohio dipped into tree foliage, sparking "a cascade of failures," the regional power grid's safeguards had failed to accommodate the disruption.[1] This electrical delivery system, in the form of the transmission and distribution grid, continues to dictate who has access to power.

Previous chapters have addressed the subject of energy through the lens of technological development, environmental impact, use, and transition, but one often overlooked theme is equity. Earlier discussions about uranium and fossil fuels have noted that those living at sites of energy extraction and production disproportionately feel the health and environmental consequences. But the following discussion will lean more toward access and the distribution of that energy, toward users, and toward the political, economic, and social power we may not realize we have in deciding who has and receives power and how we use *electrical* power, in particular.

In order to achieve an economy of scale, the development of the grid required building demand for the electrical customers, or "load," to make distribution economically feasible. A utility's economic viability is dependent on how, and how much, consumers use energy. Invented during the Gilded Age of the second Industrial Revolution and the rise of big business, electricity became a commodity for private enterprise, rather than a public service (until the 1930s). Utilities thus did not prioritize equal distribution of their product across all populations.[2]

Even museums and historic sites not explicitly related to the history of electricity can engage audiences in the topic. Places like Edison's home and laboratory in Menlo Park, New Jersey, is a national park that emphasizes themes of innovation and invention. But in fact, so many people and entities worked throughout history to develop lighting and lighting system equipment that small towns like Ansonia, Connecticut (where William Wallace's brass factory inspired Edison), to Niagara Falls, New York, can tell part of the history.[3]

Histories and Contexts

The demand for electricity roughly doubled every decade in the twentieth century, but most people initially hesitated to adopt new technology for cultural reasons. Electricity eventually transformed daily life at home and at work, moving humankind from the industrial age to one where energy is integrated into all facets of life and landscapes. Electricity allowed complete climate control over our interior environments by burning fossil fuels, which have, in turn, altered the climate outside. These changes have likewise altered the physiology of most living things over time. The success of light bulbs erased any clear division between day and night.

Prior to electric light, homes and workplaces were illuminated only by some form of flame, either candle or oil. While most histories of electricity focus on industrialists and inventors, Jeremy Zallen's *American Lucifers* examines the sacrifices of laborers, free and enslaved, in the quest for artificial light.[4] American companies led the world in developing electrical infrastructure; however, Europeans developed the first electric lights by the late nineteenth century. Italian Alessandro Volta's battery helped England's Humphry Davy and his countryman Michael Faraday conceive of how to produce electrical current by passing a magnet along a copper wire. In what became the "arc lamp," the invention of gas-filled glass tubes that sustained an electrical current between two pieces of charcoal in an arc attracted great commercial interest in street lighting. Unfortunately, it proved too bright and impractical to use inside the home. Many hoped to devise an incandescent bulb that would replace gas as a source of energy for light, especially for the indoors. Based on previous patents, American Thomas Edison, already a famous inventor with advancements in telegraphy and the phonograph, developed a filament that could hold the even, clear light from a current in a gasless, odorless, and smokeless bulb for several hours. In *Edison's Electric Light: The Art of Invention* (2010), Robert Friedel and Paul Israel mined the Menlo Park archives to describe how Edison's laboratory conceived an entire lighting system (a centralized underground network that could serve areas of a city) from idea to commercialization. Friedel and Israel credited external "expectations, confidence, and resources" for establishing the first commercial coal-fired steam engine, which supplied an electric power station at Pearl Street in Lower Manhattan. But rather than "analyzing and understanding technological change," they discussed Edison's creative process at a particular place and in a particular time.[5]

Once he successfully used a carbonized cotton thread as filament, Edison issued several licenses to various companies to utilize his incandescent light bulbs and direct current (DC) circuit equipment throughout the 1880s. However, the DC system was limited. The electrons in DC current moved in one direction from the generator to the motor or user, with the current weakening after about a mile or so. However, the electrical impulses in alternating current (AC) oscillated at back-and-forth intervals, maintaining current strength for longer distances. The rate of the alternating current is known as *frequency*. Frequency (measured in *cycles*) dictates the force of the electrical current (*voltage*) and a decrease in the amount of electricity transmitted at a time. Transformers or converters step voltage up or down through a set of electromagnetic coils.

As Edison worked to perfect the DC system, Croatia-born Serbian Nikola Tesla patented a simply constructed polyphase motor. Tesla's motor could convert AC into mechanical energy, making the system a more consistent and cheaper alternative to Edison's DC system. After Edison rejected changing his DC system to accommodate it, George Westinghouse secured the manufacturing rights from Tesla. The DC system's limited range required a power plant every mile or so and limited service to the cities. Long-distance transmission with the AC system required fewer power stations and less copper, with the ability to deliver power to more users living farther apart, which was especially important for the interior of the United States.

Therefore, in 1884, the Westinghouse Electric and Manufacturing Company became Edison's strongest competitor to the DC system by consolidating patents and companies to build its own system based on AC. The rivalry between Edison's DC and Westinghouse's AC systems contributed to the electrical industry's incongruity and consumers' fears. Edison even played up public fears that AC current was so dangerous that it powered the electric chair in prisons, labeling AC the "death current." He publicly referred to electrocution as "being Westinghoused." He even publicly "Westinghoused" domestic animals to create revulsion and fear for his rival's product. Regardless, with Tesla and his AC system, Westinghouse won contracts to power the city of Buffalo, New York, and eventually multiple generators across the Northeast through a model hydroelectric power plant at Niagara Falls. Westinghouse solidified the dominance of the AC system in the electrical market with "the Great White City" at the 1893 World's Columbian Exposition in Chicago. These advances encouraged the development of hydroelectric dams that could produce electricity far from dense population centers and send it through longer transmission lines to distant cities across the country.[6]

Interpretative programming can certainly exploit the drama around the Edison and Westinghouse rivalry, which unfolded within the historical context of innovation, the rise of big business, and the Gilded Age. Author Jill Jonnes (*Empires of Light*, 2003) wrote a gripping account, and Hollywood even made a major motion picture (*The Current War*, 2019) starring Benedict Cumberbatch and Tom Holland.[7] The story illustrates the powerful influence of not just economics, but charismatic personalities and public demand on the design of energy systems and infrastructure. Tesla himself thought beyond the AC system, promoting wireless electricity (through electromagnetic induction), as well as wind and solar energy. While Edison's DC technology is now favored for its stability in solar cells, LEDs, and electric vehicles, the AC system is the grid we depend upon today.

At the turn of the century, Edison's former private secretary left General Electric to manage utilities in Chicago. There, he consolidated the city's utilities for economy of scale. In powerhouses, he replaced steam engines with turbines. Corporate competition further encouraged technological advancement. The electrical industry did not standardize its equipment, so different systems remained incompatible and incongruent. The development of the grid, which integrates different parts and technologies, necessitated standardized equipment. As historian of technology David Nye observed, "What began as a voluntary choice became a requirement."[8] Samuel Insull's model of system consolidation increased the efficiency and affordability of regional electric utility systems.

With standardization, utilities developed "natural monopolies." According to the principle of economy of scale, the larger the grid, the more users and the lower the costs. The natural monopolies of private utilities primarily served urban areas and controlled the market until the Great Depression. Monopolies require regulation to maintain reliability, affordability, and equal distribution of power. The standardized grid creates codependency between power producers and a network of consumers to be reliable. It also distances the process of energy generation from energy use. Energy sources and the process of producing mechanical or electrical power disappears from users' sight and awareness, with the exception of the delivery system (power lines). As the public became more dependent on electrical power and private utilities based their service on profits, the federal government stepped in to ensure equal access, accusing utilities of stifling competition. That conflict became a metaphor for battling political ideologies as private utilities erroneously accused groups like the Tennessee Valley Authority (TVA) and the Rural Electrification Administration (REA) of socialism.[9]

The first historical accounts of electricity focused largely on innovation and progress. Many wrote with the assumption that technology alone, irrespective of political or social influences, inevitably drives change in people's lives (a notion known as "technological determinism"). Other accounts examined the economic aspects of the early twentieth-century "Power Wars," which pitted private power companies against government-owned systems. By the late twentieth century, historians began to argue that large, integrated technological systems actually reflected political hegemony. In *Networks of Power* (1983), Thomas Hughes convincingly argues that technological systems, and the builders of those systems, defined the modern American nation.[10] Other works from the 1990s, like Mark Rose's *Cities of Light and Heat* (1995), Ronald Tobey's *Technology as Freedom* (1995), and Jay Brigham's *Empowering the West* (1998), explored the role of politics in the mass adoption and use of technology.[11] Martin Melosi's *The Sanitary City* (2000) illustrated the rich historical influence of urban infrastructure on urban and community development.[12]

In the shadow of growing calls to deregulate the electrical industry at the dawn of the twenty-first century, historians focused upon the political and social inequities inherent in the regional and national electrical grid. Rather than the story of business and technology that Hughes tells, Andrew Needham's *Power Lines* (2014) describes regional urban growth in the Southwest on the model of William Cronon's *Nature's Metropolis: Chicago and the Great West* (1992). The Colorado Plateau became the center for power generation in the late twentieth century, with transmission lines physically tying together formerly distant and independent systems and exacerbating a political and economic power imbalance between the urban and rural (in this case Native) populations. The integration of these two landscapes merged energy policy with Indian policy, causing great challenges to concepts of environmental and social justice. The political and economic power remained at the periphery, at the site of energy consumption, rather than the plateau, the site of energy production.[13] Needham's explanation of power system development across the Southwest makes a good companion to *Routes of Power* (2016), where Christopher Jones examines the design and development of regional oil pipelines and gas lines, as well as the grid. Such histories about deep investments in infrastructure help explain how replacing fossil fuels with renewable energies remains a daunting task.[14]

David Wuebben's *Power-Lined* (2019) provides an excellent cultural analysis of the aesthetics of wired networks on the landscape and how they impact the public's perception of not just energy, but place.[15] Exploring the nuances of historical energy distribution and use through local and/or regional studies can help us negotiate our complicated generation and consumption of energy in the future. As defined by Sarah Deutsch in *No Separate Refuge* (1989), a "regional community bound by kinship as well as economy" does not always adhere to political boundaries, and likewise, electrical service areas did not always stop at state or reservation boundary lines. Power lines created vast new regional networks.[16] As I showed in *Electrifying the Rural American West: Stories of Power, People, and Place* (2009), communities designed electrical systems specific to their local and regional needs. I examined the sociocultural influence of ordinary people on the electrification process against the cultural, environmental, and political features of the American Southwest. Paul Hirt (*The Wired Northwest*, 2012) and Casey Cater (*Regenerating Dixie*, 2019) similarly studied the American Northwest and South, respectively. We all essentially argue that the electrical system, its design, and its use reflected regional characteristics. Emphasizing process and adaptation, Hirt invokes organic language to describe the electrical infrastructure as a constantly changing "web" responding to eras of crisis.[17] After the collapse of the Southern public power movement in the 1970s, Cater describes how the South eventually shifted from hydropower back to fossil fuels to meet regional energy demands in the name of progress, convenience, and consumerism, in spite of protests by environmental and consumer rights groups.

Regional traditions foreshadowed the role of an emerging electrical system, one that linked individual communities, altered the area's economy, and introduced selective urban amenities into homes. Beyond their role as individual consumers, rural residents especially enlisted electricity to exploit the land's natural resources, to create economic opportunity and stability, and so preserve existing rural communities within an emerging twentieth-century industrial society.[18]

Regional utilities marketed the new technology specifically to the region's *milieu*. For example, with racially charged language in the middle of the twentieth century, Southerners fused the "Old South" rhetoric of slavery with the Progressive era values of conservation to put the formerly "lazy" rivers to work as the region's new labor force. Utilities even featured racial stereotypes, such as having "Reddy Kilowatt" assume the role of electricity as the South's "cheapest servant." At least one prominent activist who resisted the electrical industry's actions invoked "the Lost Cause" to push back against the environmental manipulation and destruction of "traditional" southern agricultural landscapes. Nonetheless, while segregation and housing inequity limited access, urban life exposed all races and classes to the new technology via streetlights and streetcars. Likewise, across the country in the Southwest, the tribally owned utility operated by the Diné (Navajo) advocated for the benefits of electricity and appliances. They reinterpreted "Reddy Kilowatt" as "Kee Kilowatt," an anthropomorphic light bulb that introduced Diné of all ages to indoor lamps, the iron, the washing machine, the refrigerator, and the television.[19]

The earlier chapter on fossil fuels mentioned efforts by Native tribes to control their own energy development as a way to become economically independent and to achieve greater sovereignty over their own political and economic affairs. Native people accessed power for domestic and industrial use and asserted tribal sovereignty through regional distribution systems. Since the 1970s, many Native communities began to embrace the tenets of the "appropriate technology" movement to articulate ways to preserve culture and identity through local control.[20] The Diné developed their own utility (the Navajo Tribal Utility Authority), modeled on rural cooperatives, in order to secure distribution of electricity and dictate use. In *Network Sovereignty: Building the Internet Across Indian Country* (2017), Marisa Elena Duarte argues that building successful technological systems for Native communities can only achieve social and economic justice when they are place-based and reflect the environmental and cultural systems of its users. Reservations hold potential for exercising power over postindustrial, often renewable, energy sources, as well.[21]

Yet reservations across the country show far fewer electrified households than the general population despite these lands serving as a center for logging, hydroelectric dams, coal deposits, oil, and uranium mines. The rise of cities and their dependence on electricity in the postindustrial twentieth century necessarily increased urban dependence on coal, a resource associated more with nineteenth-century industrialism, for fuel. By rejecting hydroelectric dams as harmful to the environment, environmental groups, as well as Native communities additionally concerned for sacred sites, inadvertently gave rise to rural coal plants like the Navajo Generating Station near Page, Arizona. By May 1966, Pennsylvania congressman John Saylor asked the Bureau of Reclamation's Floyd Dominy to "tell us why any dam built only to generate power is ever necessary, given the ability of atomic energy or coal to generate electricity far more inexpensively."[22] Unfortunately, we can interpret coal's second life *after* the Progressive conservation era that promoted waterpower as a tragic consequence of other progressive intentions for the American West: economic independence for tribes, postwar urban development, and environmental concerns about water supply and quality.

Electrical distribution systems emerged from existing cultures and geographies that determined who had access to electricity. Thus, electrification "did not occur for everyone at the same pace nor to the same degree."[23] Ultimately, its impact depended upon a combination of local factors including location, population density, culture, residential segregation, housing type and home ownership, an area's economic and demographic makeup, and grassroots activity. When these factors denied communities electricity, people were left behind in an era where electrified homes increasingly became the standard for modern living.[24] All this background suggests several themes and topics through which to interpret the electrical grid and what the system says about how we produce and use energy.

Interpreting Electricity

In the 2000 film *Oh Brother, Where Art Thou?* George Clooney's character, Ulysses Everett McGill simply and briefly summarizes the intent and promise of a federally funded electrical system in the 1930s: "Yessir, the South is gonna change. Everything's gonna be put on electricity and run on a paying basis. Out with old spiritual mumbo-jumbo, the superstition and the backward ways. We're gonna see a brave new world where they run everyone a wire and hook us all up to the grid." As interpreters, it can be challenging to go beyond technological and economic jargon, and like Everett McGill, make this history accessible and personal to our audience. The infrastructure reflects technological innovation, but rather than stressing science and engineering, we should help visitors see the connections between the grid and their everyday experience.

Many energy-focused museums lean toward engaging and often interactive science. Even when presenting the historical development of electrical machinery in innovative and exciting ways, as at the popular SPARK Museum in Bellingham, Washington, the interpretation supports a narrative of technological progress.[25] The older *Lighting the Revolution* exhibit at the Smithsonian's National Museum of American History at least addresses "consequences" and "conservation," inviting visitors to think about their own choices for energy use, and even historicizing energy conservation itself. However, that message may be smothered by the display of early electrical material culture.[26] Interpreting the role of electricity *in* history, rather than just the history *of* electricity, can address a large swath of themes focused upon access and use.

Consumerism

Utilities heavily promoted appliance use to build load and their systems, which meant finding the most users per square mile.[27] The economic feasibility of the electrical grid greatly exacerbated an already growing socioeconomic, political, and cultural divide between urban and rural communities in the United States. Both the media and reformers increasingly associated "the inferiority of rural life" with lacking electrical power as America drifted away from Civil War sectionalism to becoming almost as divided along rural and urban lines. Through the introduction of modern technology and presumably modern ways of thinking, some saw electrification as a way to enlighten communities where religious fundamentalism and opposition to scientific theories like Darwinism had grown more entrenched. The reformers of the country life movement (c. 1900–1920) believed that access to and proper use of new technology would improve rural lives. True social democracy would only occur with equal access to technology, and these ideas would largely influence government policies toward rural America. To close this access gap, the government assumed responsibility for rural electrification first through the TVA and then the REA.[28] An REA cooperative could only be feasible if enough users created enough load (demand) to support economically sustainable systems.

Because the REA reflected the philosophies of the country life movement reformers, the agency placed immense emphasis on domestic use. The REA magazine *Rural Electrification News* claimed that electricity would rescue a rural woman from her life of drudgery. Articles went so far to predict that once a home became electrified, a woman's work would be "more like play."[29] Perhaps even more significantly, new appliances would replace the time-honored cultural and economic traditions of food preservation and clothes washing. Lighting and plumbing in rural schoolhouses supplemented the efforts of the adult educational programs by exposing children to various electrical appliances and teaching them about the new technology. At a minimum, electrical lighting allowed rural children more hours for studying. Nationwide, science and home-economics teachers incorporated electricity into their curriculum even if the students did not have it yet in their homes. These federally led and locally supported efforts hoped to curb the tendency of young people to leave the farm for the cities, raise interest in rural living opportunities, and create a demand for electricity that would support electrification systems. Like the utilities' demonstration programs in urban areas, the government designed its educational programs for white, native-born, middle-class women to work on industrialized farms. The Agricultural Extension Service formed clubs to teach rural women how to use new devices. The agency's influence on widespread electrical appliance adoption often depended on how well individual educators and agents crossed ethnic and socioeconomic lines.[30]

In his book *Consumers in the Country* (2000), historian Ronald R. Kline argues that rural communities did not embrace urbanizing infrastructure and technologies like plumbing and electricity as the reformers hoped. Rather, a "contested interaction between producers and consumers" determined the impact of "urbanizing technologies" like electricity. Rural people exercised agency and resistance in the face of the changes that others imposed, often controlling the impact a particular technology might have on their lives and incorporating the new technology into "existing cultural patterns."[31]

Kline's examination of rural consumerism complements Lizabeth Cohen's *A Consumers' Republic* (2003), which explained that various socioeconomic traditions and constructions of gender determined the influence and various new technologies, consumer goods, and appliances in the home. It helps contextualize the classic works of Katherine Jellison's *Entitled to Power* (1993) and Ruth Schwartz Cowan's *More Work for Mother* (1983) that explore the role of gender in the use of new agricultural technologies on the farm and in the home.[32] David Nye's *Electrifying America: Social Meanings of a New Technology* (1990) serves as a more people-centered history examining use. Nye argues that people ultimately decided how to incorporate electricity into existing social patterns and landscapes. When the new technologies of the industrial age first began to emerge, a technologically literate elite hoped to guide social change in a rapidly changing era. Race, class, gender, and rural status further marginalized certain groups from "insider" status.[33] Gail Cooper explained in *Air-Conditioning America* (2002) that people were slow to give up traditional warm weather rituals and accept indoor climate control.[34] Users adjust to new norms, however, and most people do not realize their energy dependence until they no longer have access to the energy.

Not everyone readily adopted and adapted to electricity. Many older rural people remained skeptical of new devices in the home, or even feared the new technology. An Arizona woman reportedly convinced herself that the current would kill off her chickens and refused to even consider electrical power. Lack of education about and experience with the new machines that challenged rural traditions may have also contributed to their reluctance. Many saw little point in purchasing the heavy equipment for what they thought would be only limited work relief. Several government-commissioned films about the REA, circa 1940, are available online to help illustrate how the government helped convince people about the value of electricity in transforming the family farm. *Bip Goes to Town* compares the electrified farm to one without, featuring various types of electric machines "to do the hard work" such as milking and to provide needed cooling to keep milk from going sour, as well as home appliances. These films are in the public domain through the National Archives, and thus easily incorporated into in-person and online exhibits.[35] Anecdotes about farmers' hesitations to adopt new technology are worth acknowledging when we consider our need to access electrical power around the clock to do our work.

Blackouts

Every day, most people take liberal access to electrical power for granted until it fails. Blackouts ironically shed "light" on our utter dependence on electricity and the electric grid for almost every facet of life: light, heating, cooking, cleaning, mass transportation, work, entertainment, etc. "Without electricity, present day life loses most of its critical infrastructure."[36] The novelty of candles and lanterns placates many of us for a few hours, after which we assume bulbs will flutter back on and we will merely have to run around the house resetting the blinking clocks. Stories of blackouts provide fruitful events to interpret and explore our dependence upon not just electricity, but the electrical grid. Such a theme easily relates to a majority of visitor experiences. Almost everyone can remember a time when the lights went out. Sites can conduct oral history projects or provide darkened rooms for reflection as simple ways to elicit how we feel, how we respond, or how a lack of electricity affects what activities we even can do, and it can prompt visitors to compare how and what people accomplished before the electric age. David Nye's *When the Lights Went Out* (2010) would provide a solid context for these personal and more local stories at a museum. Nye points out that blackouts become communal events, sometimes positive, that serve as a break and chance to reconnect with neighbors. Artificial light so defines our interior and exterior environments that time almost stops in a blackout, and memories of time in the dark often persist far longer than our more "enlightened" lives. If a blackout occurs during a time of extreme cold or heat, and our normal mechanisms for controlling our interior climate lose power, loss of electricity can be catastrophic. Some blackouts are purposeful, as for security reasons during World War II, or conservation during the 2000 California "brownouts" as deregulation disrupted supply and demand and destabilized the grid.[37] If we only notice electricity in its absence, then discussions about blackouts can initiate engagement on topics like housing equity. Reflecting on failure can refocus attention upon infrastructure as not merely technology, but as a cultural system of privilege that dictates how people live and work.

CASE STUDY: The Eastern Mining Area Transmission Line, Arizona

Oftentimes, those very "eyesores" that observers ignore tell historically significant stories about the connections between our natural environment and the activities upon which our society and economy are based. Back when I worked for a utility company, I documented an early twentieth-century long-distance hydroelectric transmission line for a Historic American Engineering Record report. I had enlisted engineer Bill and photographer James for help identifying the resources along a historic dirt road with beautiful views of central Arizona's desert mountains. As tourists carefully navigated the Apache Trail, we stopped to scan the vista for a representative tower to document. Bill playfully turned to the sightseeing tourists and remarked, "Nice towers, eh?"[38] The befuddled visitors out to enjoy the vistas likely did not even notice the transmission towers, which mark this arid landscape as one altered by hydroelectric power. Energy infrastructure is often in the background of our daily lives, sometimes deliberately hidden or camouflaged from the public eye. We also tend to ignore it, like these tourists, in search of a more inspiring landscape or purpose.[39]

Technology is physically and functionally a product of a place and its people and therefore we can interpret those systems as we would any piece of material culture. The following is just one example. Overcoming economic, cultural, and physical barriers, formerly isolated settlements developed ties to a wider regional culture and economy, often extending beyond politically defined borders. Electrical power systems physically connected urban centers to traditionally independent rural communities with transmission and distribution lines, creating new regions, technologically defined. Once energized, those systems' wires crossed streets, bridges, rural farms, mountains, valleys, and ranges, linking even isolated settlements to deliver tools, appliances, and electrical lighting. Ann Durkin Keating's contribution to the "Exploring Community History" series, *Invisible Networks: Exploring the History of Local Utilities and Public Works* (1994), still provides guidance for researching and interpreting infrastructure.[40]

When the Salt River Project, the water and power company that owned the Eastern Mining Area (EMA) Line, needed to update its electrical system, programmatic agreements in accordance with Section 106 of the National Historic Preservation Act required documentation of the original as a HAER report. HAER reports on dams, power plants, and transmission lines can offer useful interpretation and themes for energy technology. Strategies to mitigate the impact of historic resources could include preserving signature pieces of the electrical system for public interpretation, ideally in place, but it could also mean finding a place for it in a museum's outdoor space.

The EMA transmission line connects the mining towns of Globe, Superior, and Ray in eastern Arizona, which first attracted gold and silver prospectors in the 1870s. As these minerals became depleted, miners soon discovered large copper deposits beneath the silver layers. Initially, copper was primarily used to manufacture kitchen utensils and roofs. However, as industrial enterprises expanded, copper became an essential material for conducting electricity. It emerged as one of the Arizona territory's most important exports.

The construction of the Roosevelt Dam included a temporary power plant to provide the 1,200-horsepower energy needed to run machinery, operate the cement mill, and provide lighting. However, government officials soon realized the high potential for the commercial sale of hydroelectric power to the Salt River Valley. In 1906, Congress authorized the secretary of the interior to sell excess power generated at federal reclamation projects. The receipts from these power sales would help repay the cost of building more reclamation projects. A permanent hydroelectric plant began operating in June 1909.

Figure 8.1. Workers, likely Apache, raising a steel transmission tower, March 20, 1908. Walter J. Lubken, photographer. Salt River Project Archives, Phoenix, Arizona

The early EMA transmission line delivered the energy needed to produce two of Arizona's most valuable economic resources: cotton and copper. Arizona provided these essential raw materials for industrial America throughout the early twentieth century. Thus, the long-distance transmission of hydroelectric power to both farms and mines transformed the region into an area of great wealth and prosperity. Revenue from the sale of power to the mines, the Salt River Project's largest customers, funded the building of power dams that in turn brought electrical power to the Salt River Valley (see earlier chapter). Interpreting the history behind the purpose, function, and impact of the transmission line, we can learn much about the values of the conservation era and the efficiency of hydroelectric systems like this. The line allowed the production of its own conducting material (copper), enabling a large-scale government water storage project to become an efficient yet complex hydroelectric energy and irrigation system serving most of central Arizona.[41]

Artifact Spotlight

Transmission Towers

Technological innovation rarely starts from scratch and even parts of the system can reveal interpretive themes like adaptive reuse of energy infrastructure. The design of electrical transmission towers provides an excellent examination of how design reflects particular needs and circumstances. Many museums may have glass insulators in their collections that secured wires to poles and prevented power loss on utility lines during electrical transmission (see cover).

From 1902 to 1907, Edison Electric built a series of hydro plants on the Kern River to electrify Southern California. Not only was the Kern River the highest-voltage line in the nation, it was also the first in the country not supported on wooden poles. Instead, steel towers designed by a windmill company carried the conductor wires. Electrical engineers considered a variety of factors to determine the best tower design. These factors included wind, sleet, terrain, temperature, and the probability of accidents, such as cables breaking. Designers of high-voltage electrical lines first turned to windmill manufacturers, because those structures most resembled the conditions and purposes for transmission towers.

A large portion (thirty miles) of the transmission line from Roosevelt Dam's hydroelectric plant to Phoenix traveled through the desert mountain region known as the Superstition, Apache, and Pinal Mountains. There, soil conditions are often loose and rocky. The surface is extremely uneven and among the most rugged areas in central Arizona. The vegetation is typical of the Sonoran Desert and includes dense areas of palo verde and mesquite trees, amidst cholla, saguaro, prickly pear, barrel, and pincushion cactus. The terrain, wind pressure, arid weather conditions, and high voltage required the strength and stability of steel towers, rather than wooden or steel poles. The United States Wind Engine and Pump Company in Batavia, Illinois, was the primary manufacturer of the most popular American windmill, discussed in the previous chapter. Towers were sold in intervals of five feet. Braces, which ran diagonally, were specially adapted for various height requirements.[42]

Beyond materials, in this case the towers also tell us a human story about the fate of Native people after the nineteenth-century "Indian Wars." Much of the labor force at the Roosevelt Dam included Native people of Arizona, like the Apache, and it was they who constructed the initial transmission line in the Superstition Mountains, which they completed one year before the death of Geronimo (and just two decades after his initial capture).[43]

Appliances and Product Catalogs

Appliances are artifacts that many museums can enlist from collections to discuss energy use through the lens of gender, and in some cases, culture. They serve as the final destination for all the energy pulsing through the grid. The same energy systems that powered manufacturing also industrialized the home and thus some of the interpretation discussed here about energy at sites of industry can also be part of historic house interpretation.[44] Edison's incandescent bulb was part of a household system of sockets, wiring, meters, cables, and generators. Ruth Schwartz Cowan interprets the home as an extension of the industrial system. Its operation is almost completely dependent on its integration with the electrical grid and the economic supply pipelines. All this was based on the assumption that adult women could be home, day and night. She explains, "The notion that women would do housework was almost literally cast in concrete—or, rather, in brass pipes and copper wires."[45] By 1920, most new homes came prewired for appliances like toasters,

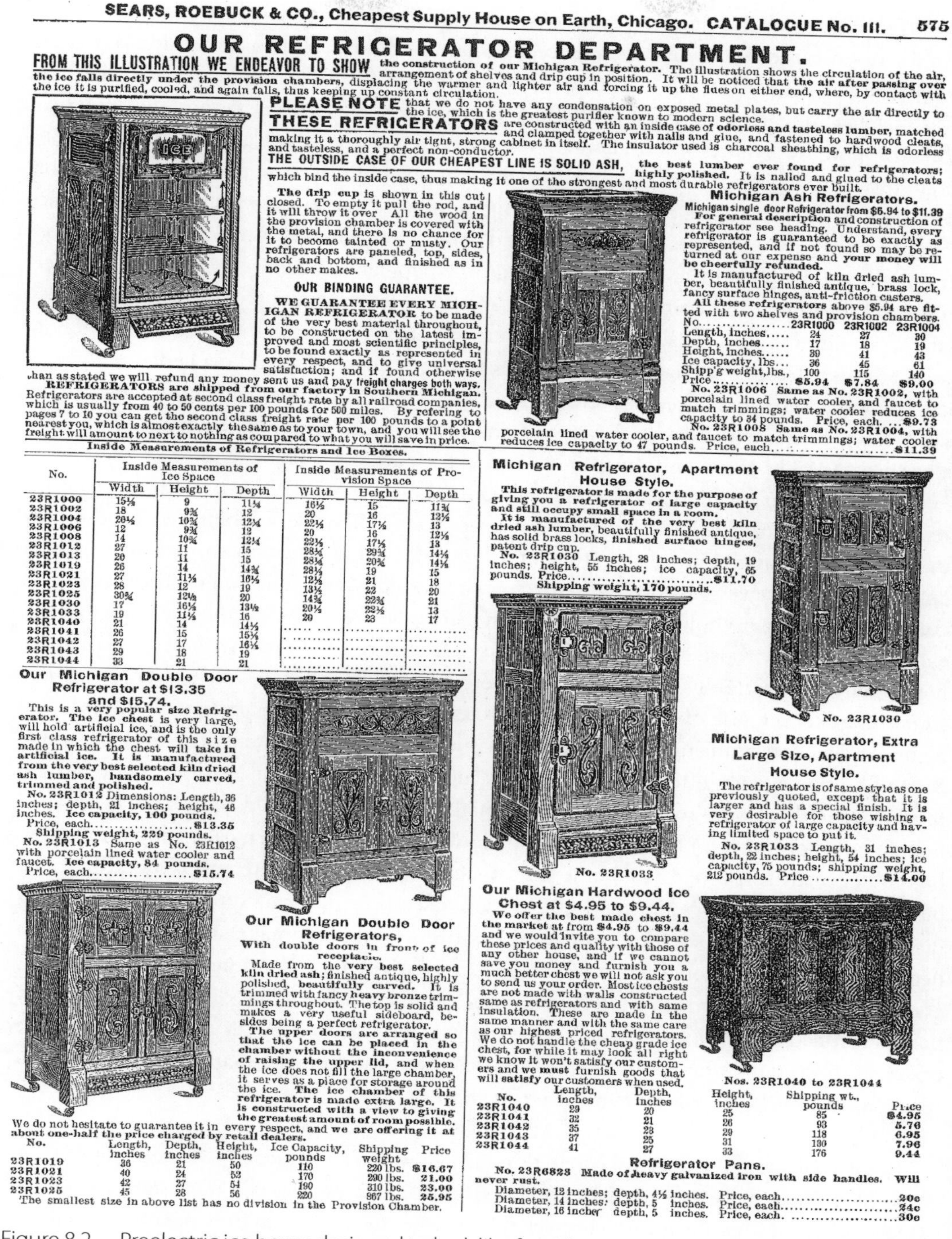

SEARS, ROEBUCK & CO., Cheapest Supply House on Earth, Chicago. CATALOGUE No. 111. 575

OUR REFRIGERATOR DEPARTMENT.

FROM THIS ILLUSTRATION WE ENDEAVOR TO SHOW **the construction of our Michigan Refrigerator.** The illustration shows the circulation of the air, arrangement of shelves and drip cup in position. It will be noticed that **the air after passing over the ice falls directly under the provision chambers,** displacing the warmer and lighter air and forcing it up the flues on either end, where, by contact with the ice it is purified, cooled, and again falls, thus keeping up constant circulation.

PLEASE NOTE that we do not have any condensation on exposed metal plates, but carry the air directly to the ice, which is the greatest purifier known to modern science.

THESE REFRIGERATORS are constructed with an inside case of **odorless and tasteless lumber,** matched and clamped together with nails and glue, and fastened to hardwood cleats, making it a thoroughly air tight, strong cabinet in itself. The insulator used is charcoal sheathing, which is odorless and tasteless, and a perfect non-conductor.

THE OUTSIDE CASE OF OUR CHEAPEST LINE IS SOLID ASH, **the best lumber ever found for refrigerators; highly polished.** It is nailed and glued to the cleats which bind the inside case, thus making it one of the strongest and most durable refrigerators ever built.

The drip cup is shown in this cut closed. To empty it pull the rod, and it will throw it over All the wood in the provision chamber is covered with the metal, and there is no chance for it to become tainted or musty. Our refrigerators are paneled, top, sides, back and bottom, and finished as in no other makes.

OUR BINDING GUARANTEE.

WE GUARANTEE EVERY MICHIGAN REFRIGERATOR to be made of the very best material throughout, to be constructed on the latest improved and most scientific principles, to be found exactly as represented in every respect, and to give universal satisfaction; and if found otherwise than as stated we will refund any money sent us and pay **freight charges both ways.**

REFRIGERATORS are shipped from our factory in Southern Michigan. Refrigerators are accepted at second class freight rate by all railroad companies, which is usually from 40 to 50 cents per 100 pounds for 500 miles. By refering to pages 7 to 10 you can get the second class freight rate per 100 pounds to a point nearest you, which is almost exactly the same as to your town, and you will see the freight will amount to next to nothing as compared to what you will save in price.

Michigan Ash Refrigerators.

Michigan single door Refrigerator from $5.94 to $11.39

For general description and construction of refrigerator see heading. Understand, every refrigerator is guaranteed to be exactly as represented, and if not found so may be returned at our expense and **your money will be cheerfully refunded.**

It is manufactured of kiln dried ash lumber, beautifully finished antique, brass lock, fancy surface hinges, anti-friction casters.

All these refrigerators above $5.94 are fitted with two shelves and provision chambers.

No.	23R1000	23R1002	23R1004
Length, inches	24	27	30
Depth, inches	17	18	19
Height, inches	39	41	43
Ice capacity, lbs.	36	45	61
Shipp'g weight, lbs.	100	115	140
Price	$5.94	$7.84	$9.00

No. 23R1006 Same as No. 23R1002, with porcelain lined water cooler, and faucet to match trimmings; water cooler reduces ice capacity to 34 pounds. Price, each. ... **$9.73**

No. 23R1008 Same as No. 23R1004, with porcelain lined water cooler, and faucet to match trimmings; water cooler reduces ice capacity to 47 pounds. Price, each ... **$11.39**

Inside Measurements of Refrigerators and Ice Boxes.

No.	Inside Measurements of Ice Space			Inside Measurements of Provision Space		
	Width	Height	Depth	Width	Height	Depth
23R1000	15½	9	11¼	16½	15	11¾
23R1002	18	9¾	12	20	16	12½
23R1004	20½	10¾	12¼	22½	17½	13
23R1006	12	9¾	12	20	16	12½
23R1008	14	10¾	12¼	22½	17½	13
23R1012	27	11	15	28¼	29¾	14½
23R1013	20	11	15	28¼	20¾	14½
23R1019	26	14	14¾	28½	19	15
23R1021	27	11½	16½	12½	21	18
23R1023	28	12	19	13½	22	20
23R1025	30¾	12½	20	14¾	22¾	21
23R1030	17	16½	13½	20½	22½	13
23R1033	19	11½	16	20	23	17
23R1040	21	14	14½	......	......	......
23R1041	26	15	15½	......	......	......
23R1042	27	17	16½	......	......	......
23R1043	29	18	19	......	......	......
23R1044	33	21	21	......	......	......

Our Michigan Double Door Refrigerator at $13.35 and $15.74.

This is a **very popular size Refrigerator. The ice chest** is very large, will hold artificial ice, and is the only first class refrigerator of this size made in which the chest **will take in artificial ice. It is manufactured from the very best selected kiln dried ash lumber, handsomely carved, trimmed and polished.**

No. 23R1012 Dimensions: Length, 36 inches; depth, 21 inches; height, 46 inches. **Ice capacity, 100 pounds.**

Price, each ... **$13.35**

Shipping weight, 229 pounds.

No. 23R1013 Same as No. 23R1012 with porcelain lined water cooler and faucet. **Ice capacity, 84 pounds.**

Price, each ... **$15.74**

Our Michigan Double Door Refrigerators,

With double doors in front of ice receptacle.

Made from the **very best selected kiln dried ash;** finished antique, highly polished, **beautifully carved.** It is trimmed with fancy **heavy bronze trimmings** throughout. The top is solid and makes a very useful sideboard, besides being a perfect refrigerator.

The upper doors are arranged so that the ice can be placed in the chamber without the inconvenience of raising the upper lid, and when the ice does not fill the large chamber, it serves as a place for storage around the ice. **The ice chamber of this refrigerator is made extra large. It is constructed with a view to giving the greatest amount of room possible.** We do not hesitate to guarantee it in every respect, and **we are offering it at about one-half the price charged by retail dealers.**

No.	Length, inches	Depth, inches	Height, inches	Ice Capacity, pounds	Shipping weight	Price
23R1019	36	21	50	110	220 lbs.	$16.67
23R1021	40	24	52	170	290 lbs.	21.00
23R1023	42	27	54	190	310 lbs.	23.00
23R1025	45	28	56	220	967 lbs.	25.95

The smallest size in above list has no division in the Provision Chamber.

Michigan Refrigerator, Apartment House Style.

This refrigerator is made for the purpose of giving you a refrigerator of large capacity and still occupy small space in a room.

It is manufactured of the very best kiln dried ash lumber, beautifully finished antique, has solid brass locks, **finished surface hinges,** patent drip cup.

No. 23R1030 Length, 28 inches; depth, 19 inches; height, 55 inches; ice capacity, 65 pounds. Price ... **$11.70**

Shipping weight, 170 pounds.

No. 23R1030

Michigan Refrigerator, Extra Large Size, Apartment House Style.

The refrigerator is of same style as one previously quoted, except that it is larger and has a special finish. It is very desirable for those wishing a refrigerator of large capacity and having limited space to put it.

No. 23R1033 Length, 31 inches; depth, 22 inches; height, 54 inches; ice capacity, 75 pounds; shipping weight, 212 pounds. Price ... **$14.00**

No. 23R1033

Our Michigan Hardwood Ice Chest at $4.95 to $9.44.

We offer the best made chest in the market at from **$4.95** to **$9.44** and we would invite you to compare these prices and quality with those of any other house, and if we cannot save you money and furnish you a much better chest we will not ask you to send us your order. Most ice chests are not made with walls constructed same as refrigerators and with same insulation. These are made in the same manner and with the same care as our highest priced refrigerators. We do not handle the cheap grade ice chest, for while it may look all right we know it won't satisfy our customers and we **must** furnish goods that will **satisfy** our customers when used.

Nos. 23R1040 to 23R1044

No.	Length, inches	Depth, inches	Height, inches	Shipping wt., pounds	Price
23R1040	29	20	25	85	$4.95
23R1041	32	21	26	93	5.76
23R1042	35	23	29	118	6.95
23R1043	37	25	31	130	7.96
23R1044	41	27	33	176	9.44

Refrigerator Pans.

No. 23R6823 Made of heavy galvanized iron with side handles. Will never rust.

Diameter, 12 inches; depth, 4½ inches. Price, each ... 20c
Diameter, 14 inches; depth, 5 inches. Price, each ... 24c
Diameter, 16 inches; depth, 5 inches. Price, each ... 30c

Figure 8.2. Preelectric ice boxes designed to look like furnishings over appliances. *The 1902 Edition of the Sears Roebuck Catalog* (New York: Bounty Books, 1969)

irons, sewing machines, vacuum cleaners, washing machines, and refrigerators, as well as lights.

We can explore this gendered understanding of labor through the lens of consumerism. As discussed earlier, individual preference, regional conditions, and cultural traditions could impact a person's reasons for adopting the most coveted electrical appliances like the iron, washing machine, and refrigerator. Exhibit displays can layer marketing and consumerism directly upon the artifacts, not to mention the REA's reports on which appliances would generate the desired electrical load to justify delivering power in the first place. Advertisements promised labor savings and prestige. Appealing to targeted audiences with purchasing choice can be an effective way to reach consumers and popularize products. Retail catalogs like those from Montgomery Ward or Sears, as well as some popular magazines and utilities newsletters, offer illustrated and written descriptions about key features of appliances with persuasive language and promises. These documents can reveal often unnoticed manipulation of gender roles or illustrate the material products that encouraged energy use and dependence. We still use some of these appliances, while others have less appeal for modern consumers. At the same time, consumers did not always use and incorporate appliances into their homes as advertised. Oral histories and the visitor's own experience can challenge intent.[46]

Trade catalogs, patents, and technical journals can provide the background on the technology, design, and intended purpose of appliances like the washing machine, windmills, heating equipment, refrigerators, and lighting.[47] Side by side comparisons of older appliances with newer versions, as artifacts or in catalogs, can also be an effective way of conveying consumer choices about energy efficiency. Many museums display appliances in exhibits, especially those with a STEM mission. A chance to interact with older appliances hands-on would certainly help someone understand exactly what electric power brought to something like the task of ironing. We need to take care to resist a narrative of progress and technological determinism that situates the issue of use out of context and loses the influence of human agency, often filtered through cultural practices. One example is the electric refrigerator. It not only replaced the traditional root cellar, but cold food storage allowed families to shop for food far less often and to offer more choice in meals, which no longer had to be locally sourced. At the same time, the appliance also threatened cultural traditions built around food preservation and preparation. Electric lights and appliances increasingly distinguished the domestic duties of urbanites from rural homemakers. With the arid, hot desert climate and seasonal crops of southeastern Arizona, rural women employed a variety of traditional customs to prepare and preserve foodstuffs. In lieu of a root cellar, many people dripped a pan of water over a screened wooden box to keep its contents cool. These practices, born out of necessity, evolved into cultural traditions and cultural products, as I recorded in my 2009 book on rural electrification:

Rosalia Salazar and Esperanza Montoya recalled how their traditionally Mexican families canned jellies and preserves. They also dried fruit, chile, and meat on wooden planks or steel wire, covering them with cheesecloth to protect the food from insects. Potatoes and carrots stayed fresh in burlap bags stored in the shade. Salazar's mother prevented meat from spoiling by frying it up, rendering its fat, and storing it in large cans. These techniques produced many traditional meals including *carne seca, carne adobada, carne asada, chile con carne*, *quesadillas*, *tortillas*, soup, rice pudding, and even fried chicken on a wood stove. European-Americans similarly dried their meat into jerky.[48]

Appliances like the electric iron and washing machine made a difference in the nature of housework. Electric irons were considerably lighter weight than the old irons, and they kept a consistent temperature without scorching clothes (or hands). Electric washing machines alleviated what was once a cumbersome, full-day activity. Laundry was also a full family activity, with some members gathering firewood to boil the clothes in water that others hauled from a spring, river, or well (if there was not a drought). After scrubbing the clothes on a ribbed board, they would only need to rinse multiple times, drying everything by the heat of the sun. In the past and perhaps as a message for the future, many chores required users to adapt to when the water was flowing, when the sun was shining, and when

Figure 8.3.   This caption reveals government sentiment toward rural electrification: *General planning. This photograph is included in the series as a vivid document on the impingement of Twentieth Century technology upon the neglected and backward rural scene. The meter on the wall of the rural shack indicates that it now receives its share of electricity from the power carried overland by the huge TVA Tennessee Valley Authority transmission line. TVA program must resolve the conflict between modern and ancient ways of life so that individuals, similar to those which are shown in the picture, will be benefited.* United States Tennessee Valley, None. Between 1933 and 1945. Farm Security Administration - Office of War Information photograph collection. Library of Congress, Prints and Photographs Division Washington, D.C. 20540 USA.

the wind was blowing.[49] However, quite notably, with appliances the burden of housework shifted almost entirely to women, creating "more work for mother." Schwartz Cowan notes, "Modern labor-saving devices eliminated drudgery, not labor."[50]

While not the kind of work required to live, both societal and cultural expectations mandated these chores. To his daughters, Epimenio Salazar would recite the proverb, "It is not a sin to be poor, but to be dirty, heaven forbid!" Entire Mexican families like the Salazars washed and ironed for the non-Mexican settlers who found the chores too burdensome. First, they gathered wood for a large fire to boil water, which they hauled up from a well or river. They boiled the clothes, scrubbed them against a ribbed board, and rinsed everything twice. After such rigorous treatment, ironing gained considerably more importance than it has today. Like many rural families, the Salazars heated a heavy cast-iron sad iron on the stove. Both its weight and heat demanded that one hold the handle with a rag or risk a severe burn. A housewife's pride, claimed Cherrel Batty Weech of the Gila River Valley, was a husband wearing a wrinkle-free shirt, even in the field. Weech recalls that her mother hated ironing by the hot stove in the summer heat so much that her family adopted a gas-powered iron in addition to several other gas or kerosene appliances (a refrigerator and washer), even though they feared the safety and reliability of gas-powered motors. Children like Weech had to trim the wicks and fill the lamps and tanks every week. Lillie Harrington of Cochise complained about chopping and hauling the wood for the stove and fireplace.[51]

Navajo Rugs (or other traditional crafts)

Although the electrical distribution system has not served the majority of those on reservations, many rural systems had Native consumers. Electricity transformed life, but people integrated electricity into existing cultural practices to transition to the industrial economy. The intensive resource development on or near reservations, especially after World War II, forced Native communities and individuals to make choices about how to incorporate electrical technology into daily traditions, cultures, and economies even as they protested oil drilling, pipelines, pollution, and lack of nuclear waste cleanup on and across their homelands. Even today, while many Native lands are sites of energy extraction, large numbers of those living on reservations use kerosene, propane, and firewood for light and heat. Electricity threatened economic dependence, but Native communities have proved adaptable and resilient in terms of incorporating electricity and electrical devices when they had access. They have at times integrated energy technology into the most traditional activities from the electrified sweat lodge to traditional methods of production for food, materials, and art. Appliances like the television or the radio could undermine storytelling traditions, but also improve communication across the vast distances of rural reservations. Electric light bulbs illuminated evening ceremonies. Microphones improved audio at the same events.[52] While electricity still challenged traditional life, Native art illustrates how Native people adapted to the changes of the non-Native market, and of electricity, by augmenting traditions.

By the 1930s, New Deal programs tried to address the limited economic opportunities that war, treaties, and assimilation programs had imposed on Native people by confining them to reservations. While sometimes these areas corresponded with homelands, reservations limited mobility to lands considered unfarmable, leaving tribes very little opportunity

Figure 8.4.   Navajo artist, Navajo Storm Rug, early twentieth century. Cowan's: A Hin-damn Company (public domain), https://www.cowanauctions.com/lot/navajo-storm-pattern-weaving-rug-3213113.

for economic independence, and often extinguishing their traditional lifeways, like hunting. The establishment of the Indian Arts and Crafts Board encouraged the production and sale of Native art, including baskets, painting, jewelry, and rugs and protected the integrity of profits of Native artisans' products for sale in the marketplace. This helped raise prices by elevating the "artifact" to "fine art." Even today, sale of "Indian Art" is one way in which tribes engage in the market economy. Production could help elevate the status of Native artisan and determine profits, and with electric light from either power lines or solar panels, some—for example, Diné weavers—could work regardless of sunlight.

Traditional rugs can provide an interesting way to examine how rural people—Native people, in particular—adapted to electricity. Designs, especially pictorial ones, began to reflect "pre" and "post" electric categories.[53] For some, electrification and modernization triggered a desire to revive older forms and styles. Some reinterpreted. While weavers have depicted lightning electricity as zigzag patterns in the past, they and other Native artists also sometimes create pictorial designs to depict contemporary reservation life. This includes things like trains, trucks, windmills, and occasionally telegraph or power lines. Electric tools have also helped refine the process of handweaving. For example, electric hair shavers smooth rug surfaces.[54]

In 2018, the Navajo Nation's Department of Natural Resources unveiled a seal that aimed to balance economic, environmental, social, and cultural values through land, water, quality of life, and power (depicted with a power line serving a hogan).[55] The design conveys the balance of tradition and modernization through the development and use of natural resources. The electrical system integrates it all, bringing the power generated by turbines and dynamos from sometimes multiple energy sources to users. That same system supplies the energy to operate gas, transportation, security, and private home systems. Planning this infrastructure unfortunately reflects spatial politics and political power as much as technological or topographic feasibility.

Those with access to plentiful, reliable, and safe electrical power are most often those with access to economic and political power. We are just beginning to understand how these systems shape behavior and ideas about energy for decades. The more we understand about the decisions behind the design of our energy infrastructure, the more we can understand how to design infrastructure for newer, cleaner, and renewable energy technologies in a way that equitably distributes electrical and political power.

Notes

1. Leah S. Glaser, *Electrifying the Rural American West: Stories of Power, People, and Place* (Lincoln: University of Nebraska Press, 2009); J. R. Minkel, "The 2003 Northeast Blackout—5 Years Later," *Scientific American*, August 13, 2008, accessed January 26, 2022, https://www.scientificamerican.com/article/2003-blackout-five-years-later.
2. "Load" is a term used to describe various types of stresses on a structure, including electrical voltage, physical pull, wind pressure, and climate, but also (relevant to this discussion) the power demand. David E. Nye, *When the Lights Went Out: A History of Blackouts in America* (Cambridge, MA: MIT Press, 2010), 20; David Nye, *Consuming Energy: A Social History of American Energies* (Cambridge, MA: MIT Press, reprint, 1999), 139–41.

3. Robert Friedel and Paul Israel, *Edison's Electric Light: The Art of Invention* (Baltimore: Johns Hopkins University Press, 2010), 4–5.

4. See Jeremy Zallen, *American Lucifers: The Dark History of Artificial Light, 1750–1865* (Chapel Hill: University of North Carolina Press, 2019).

5. Friedel and Israel, *Edison's Electric Light*, 189–200; Department of Energy, "The History of the Lightbulb," November 22, 2013, accessed January 27, 2022, https://www.energy.gov/articles/history-light-bulb.

6. Jill Jonnes, *Empires of Light: Edison, Westinghouse, and the Race to Electrify the World* (New York: Random House, 2004).

7. Jonnes; Alfonso Gomez-Rejon, Dir, *The Current War*, film, 101 Studios, 2019.

8. Nye, *When the Lights Went Out*, 17.

9. Glaser, 36–40, 109–16.

10. Merritt Roe Smith and Leo Marx, eds., *Does Technology Drive History? The Dilemma of Technological Determinism* (Cambridge, MA: MIT Press, 1994); Donald Worster, *Rivers of Empire: Water, Aridity, and the Growth of the American West* (New York: Oxford University Press, 1985); Thomas Parke Hughes, *Networks of Power: Electrification in Western Society*, 1880–1930 (Baltimore: Johns Hopkins University Press, 1983).

11. Ronald Tobey, *Technology as Freedom: The New Deal and the Electrical Modernization of the American Home* (Berkeley: University of California Press, 1995); Jay Brigham, *Empowering the West: Electrical Politics Before FDR* (Lawrence: University of Kansas Press, 1998); Mark H. Rose, *Cities of Light and Heat: Domesticating Gas and Electricity in Urban America* (University Park: Pennsylvania State University Press, 1995).

12. Martin V. Melosi, *The Sanitary City: Urban Infrastructure in America from Colonial Times to the Present* (Baltimore: Johns Hopkins University Press, 2000).

13. T. Andrew Needham, *Power Lines: Phoenix and the Making of the Modern Southwest, 1945–1975* (Princeton, NJ: Princeton University Press, 2014).

14. Needham, *Power Lines*; Christopher Jones, *Routes of Power: Energy and Modern America* (Cambridge, MA: Harvard University Press, 2016).

15. David Wuebben, *Power-Lined: Electricity, Landscape, and the American Mind* (Lincoln: University of Nebraska Press, 2019).

16. Sarah Deutsch, *No Separate Refuge: Culture, Class, and Gender on an Anglo-Hispanic Frontier in the American Southwest, 1880–1940* (New York: Oxford University Press, 1987), 9–10.

17. Paul Hirt, *The Wired Northwest: The History of Electric Power, 1870s–1970s* (Lawrence: University of Kansas Press, 2012).

18. Glaser, 20; Robert Kline, *Consumers in the Country*.

19. Casey P. Cater, *Regenerating Dixie: Electric Energy and the Modern South* (Pittsburgh: University of Pittsburgh Press, 2019); Glaser, *Electrifying the Rural American West*.

20. The American Indian Science and Engineering Society (AISES) published the quarterly magazine *Winds of Change* to convey this adaptability.

21. See Marjane Ambler, *Breaking the Iron Bonds: Indian Control of Energy Development* (Lawrence: University of Kansas Press, 1990).

22. Needham, *Power Lines*, 201.

23. Paula M. Nelson, "Rural Life and Social Change in the Modern West," in Douglas Hurt, *The Rural West Since World War II* (Lawrence: University of Kansas Press, 1998), 39.

24. See Tobey, *Technology as Freedom*, and George Lipsitz, *The Possessive Investment in Whiteness: How White People Profit from Identity Politics* (Philadelphia: Temple University Press, 1998), 6–8.

25. For example, the General Electric Company Museum of Innovation and Science, Schenectady, New York, and the SPARK Museum of Electrical Invention, Bellingham, Washington.

26. *Lighting the Revolution*, National Museum of American History, Smithsonian Institution, Washington, D.C. Also accessed January 5, 2022, https://americanhistory.si.edu/lighting/20thcent/consq20.htm.

27. See Richard Hirsch, *Technology and Transformation in the American Electric Utility Industry* (Cambridge, MA: MIT Press, 1989).

28. Ronald R. Kline, *Consumers in the Country: Technology and Social Change in Rural America* (Baltimore, MD: Johns Hopkins University Press, 2002); Righter, 59–125; The Country Life Commission's findings that rural Americans lacked access to the electrical grid parallels how today's digital divide affects social and economic inequality in the rural West, particularly when such services are not regulated at the federal level.

29. Barbara Rudd, "Woman's Work is More Like Play on the REA Electrified Farm," *Rural Electrification News* 2:3 (November 1936), 10–11.

30. Glaser, 65–67; Amy Elisa Ross, "Every Home a Laboratory," 3, 52–53; "Crossing Ethnic Barriers in the Southwest: Women's Agricultural Extension Education, 1914–1940," in Joan M. Jensen, *Promise to the Land: Essays on Rural Women* (Albuquerque: University of New Mexico Press, 1991), 220–30; Katherine Jellison, *Entitled to Power: Farm Women and Technology, 1913–1963* (Chapel Hill: University of North Carolina Press, 1993), 9–10.

31. Kline, *Consumers in the Country*, 6, 10, 19, 131.

32. Brigham, *Empowering the West*; Tobey, *Technology as Freedom*; Ruth Schwartz Cowan, *More Work for Mother: The Ironies of Household Technology from the Open Hearth to the Microwave* (New York: Basic Books, 1983); Mark Rose, *Cities of Light and Heat*; Jellison, *Entitled to Power*; Lizabeth Cohen, *A Consumers' Republic: The Politics of Mass Consumption in Postwar America* (New York: Alfred A. Knopf, 2003).

33. David Nye, *Electrifying America: Social Meanings of a New Technology 1880–1940* (Cambridge, MA: The MIT Press, 1990); Carolyn Marvin, *When Old Technologies Were New: Thinking About Electric Communication in the Late Nineteenth Century* (New York: Oxford University Press, 1990).

34. Gail Cooper, *Air-Conditioning America: Engineers and the Controlled Environment, 1900–1960* (Baltimore: Johns Hopkins University Press, 1998).

35. Katherine Benton-Cohen, "Common Purposes, Worlds Apart: Mexican-American, Mormon, and Midwestern Women Homesteaders in Cochise County, Arizona," *Western Historical Quarterly* 36:4 (Winter 2005), 435; Weech, interview; Nelson Peck, interview with author, Pima, AZ, August 7, 2000; Pare Lorentz and Joris Ivens "Rural Electrification in Ohio," "Power and the Land," Rural Electrification Administration, U.S. Department of Agriculture by Rural Electrification Administration for the U.S. Department of Agriculture, 1940, 1941, https://archive.org/details/RuralElectrification; infinityfilmarchive, *Bip Goes to Town*, 1941, *YouTube*, https://www.youtube.com/watch?v=HhOaMm6Zqic, posted July 20, 2008.

36. Nye, *When the Lights Went Out*, 182.

37. See Nye, *When the Lights Went Out*.

38. Leah S. Glaser, "Nice Towers, eh? Evaluating a Transmission Line in Arizona," *CRM: Cultural Resource Management* 20:14, U.S. Department of the Interior, National Park Service (1997): 23–24.

39. John A. Kouwenhoven, *The Beer Can by the Highway: Essays on What's "American" about America* (New York: Doubleday and Company), 1961.

40. Ann Durkin Keating, *Invisible Networks: Exploring the History of Local Utilities and Public Works* (Malabar, FL: Krieger Publishing, 1994).

41. Leah S. Glaser, "The EMA Transmission Line," No. AZ-6-B. Historic American Engineering Record (HAER), National Park Service, Western Region, 1996.

42. T. Lindsay Baker, *A Field Guide to American Windmills* (Norman: University of Oklahoma Press, 1985), 7–10, 22, 70; Leah S. Glaser, "The EMA Transmission Line," 1996.

43. *Arizona Republican*, 26 July 1908.

44. Cowan, 6.

45. Cowan, 91–93, 101.

46. Cowan, 218–19.

47. Lawrence B. Romaine, *A Guide to American Trade Catalogs, 1744–1900* (New York: Dover Publications, 1990).

48. Glaser, *Electrifying the Rural American West*, 21–22.

49. Glaser, 22.

50. Cowan, 91–93, 101.

51. Glaser, 21–22.

52. Glaser, "An Absolute Paragon of Paradoxes"; "Native American Power and the Electrification of Arizona's Indian Reservations," *Indians and Energy: Exploitation and Opportunity in the American Southwest* (Santa Fe, NM: School of Advanced Research, 2010), 161–62, 186–87.

53. Glaser, "An Absolute Paragon of Paradoxes," 186.

54. Andrew Nagen, "Collectors' Guide," accessed January 28, 2022, https://www.collectorsguide.com/fa/fa085.shtml; John Brandenburg, "Living Without Conveniences Part of Navajo Rug Weaving," *Oklahoman* (November 6, 1981), accessed January 28, 2022, https://www.oklahoman.com/story/news/1981/11/06/living-without-conveniences-part-of-navajo-rug-weaving/60364391007; Lorraine Tso, "Train Blanket Revival," Garland's Navajo Rugs, accessed January 28, 2022, https://www.garlandsjewelry.com/products/train-blanket-revival.

55. See Ann Lane Hedlund, *Navajo Weaving in the Late Twentieth Century: Kin, Community, and Collectors* (Tucson: University of Arizona Press, 2004).

CHAPTER 9

Conclusion

Some Thoughts on Best Practices

THIS CONTRIBUTION TO the *Interpreting History* series only scratches the surface in its review of the literature and the current practices around the theme of energy. As I completed this manuscript in 2022, the U.S. Congress separated its infrastructure bill from what was intended to be companion legislation addressing climate change. The separation of this legislation shows a clear cognitive disconnect between energy, the environment, infrastructure, and equity, when the history of energy shows how integrated it all is. A few months later, on June 30, the Supreme Court ruled that the Environmental Protection Agency could not enforce emissions standards. Without the aid of experts to balance energy use with environmental health, public understanding and the public will for change is critical. The National Park Service and the National Association for Interpretation stress that effective interpretation connects tangible objects to intangible, ideally universal, concepts.[1] The material culture, historic resources, and sites of energy production, distribution, and use are tangible artifacts that can provide the public with proximity to those energy processes upon which daily modern life depends.

Furthermore, historical thinking and understanding can move the public beyond data and statistics. Historical methodology can examine change over time and analyze the actions and attitudes toward unsustainable extraction of natural resources within racial, social, and economic contexts. Even the study of renewable energies like wind and solar likewise illuminates the tensions between historical practices and a sustainable future. These tangible examples provide remarkable opportunities for conveying *intangible* and *universal* ideas about how we as humans have developed and used energy, and how those uses have reflected and influenced our culture and our values.

Public history can engage new audiences to think about energy use and production beyond the popular nostalgic activities and trips down memory lane.[2] As Michelle Moon and Cathy Stanton have observed about the food chain, much of the general public is disconnected from the production end, even though most of us are avid consumers of both food and energy (and food for our energy). Few of us understand how either one reaches us, and without recognizing those connections, changing behavior and adapting to new energy technologies and energy systems will take far more time than we have.

In addition to technological development, energy history has evolved around social, cultural, and economic choices. We know now that such decisions have directly contributed to global warming. Environmental historians have written prolifically over the last twenty years about the social, economic, and cultural explanations and documented the decisions that led to excessive fuel consumption and its toll on the environment. They have analyzed, challenged, and reinterpreted commonly held myths. Beginning around the 1990s, public historians began exploring the ways the field overlapped with environmental history. Increasingly, public historians have embraced and encouraged programming and interpretation around issues of civic engagement and social justice. Programming, interpretation, and new approaches to storytelling that encourage visitors to likewise explore environmental justice and sustainability issues must become part of that trajectory.[3]

Museums and historic sites can, and really must, participate in the fraught national conversation on climate change and environmental sustainability through interpretation, shared authority, collaborative partnerships, and historic resource preservation and management. This can, but does not have to, entail a dramatic reinterpretation, but it should involve reflexive practice. Transparency about why energy is important to a cultural institution's history and mission might be a necessary strategy for enlisting stakeholders in the process and the topic. The discussions in this volume have aimed to explain the evolution of some best practices that can help even smaller institutions take initial steps. These ideas include: 1) enlisting techniques borrowed from current best practices in the field of public history for engaging the public in activism and approaching energy interpretation as an urgent and controversial topic, worthy of hard conversations about not only our "difficult history" regarding energy, but what will certainly be a difficult future and require difficult decisions; 2) identifying and maintaining popular visitor experiences, but contextualizing the popular appeal of those attractions as part of the evolving interpretation. Adding adjacent or supplemental narratives to exhibits or through public programs can help emphasize energy transitions over time and into the future; 3) identifying the relationship of the natural resource to the energy technology, which often dominates interpretation of historic energy-related resources; 4) sharing stewardship over the narrative with local community partners and connecting to local stories; 5) developing programming to preserve trade knowledge and skills of practices like charcoal burning and steam engine operation; and 6) threading the needle between environmental stewardship and historic preservation.[4]

Some historic processes and resources, such as those dependent on fossil fuels like steam trains and ships and power stations, should not continue to operate as they have been, even in an educational context. We need to find other ways to preserve these through written and digital documentation, selective operation, and preservation in place. We need to find other ways to interpret collections, but we also need not allow collections to completely define and limit the content.

Volatile politics, the epidemic of disinformation, and a rapidly changing planet challenge stewards of historic places and resources who must satisfy donors and supporters, but also educate and expand visitorship. When museums and historic sites focus only on energy stories of technological progress, celebration, or even tragedy, we miss that important piece of interpretation that Tilden identified sixty-five years ago: provocation. How we interpret history matters, and thinking of history as anything other than a dynamic and nuanced story that can inform present and future knowledge, values, and behavior, is at worst, neglectful and reckless, and at best irresponsible. Hopefully this book helps site managers and museum professionals to at least begin to temper the challenges of today's climate, both literal and political. Ideally, by contextualizing energy choices and preserving energy resources, historic places can make energy more visible and comprehensible for visitors, and they can help inspire the public to insist upon energy policies that preserve the future.

Notes

1. Lisa Brochu and Tim Merriman, *Personal Interpretation: Connecting Your Audience to Heritage Resources*, 3rd ed. (Ft. Collins, CO: InterpPress, 2015); David L. Larsen, ed., *Meaningful Interpretation: How to Connect Hearts and Minds to Places, Objects, and Other Resources* (Fort Washington, PA: Eastern National, 2003).
2. Michelle Moon, *Interpreting Food at Museums and Historic Sites* (Lanham, MD: Rowman and Littlefield, 2015); Michelle Moon and Cathy Stanton, *Public History and the Food Movement: Adding the Missing Ingredient* (New York: Routledge, 2018).
3. Andy Kirk, Joseph Cialdella, Leah S. Glaser, Melinda Jetté, Nancy Germano, and William Ippen, "Public Education and Environmental Sustainability Best Practices and Resources" (draft, Spring 2018), https://ncph.org/phc/public-history-education-and-environmental-sustainability-2017-working-group/public-history-education-and-environmental-sustainability-best-practices-and-resources-draft.
4. The Historic Preservation Fund, established in 1977 to provide financial assistance to states, relies on oil and gas lease revenues on the Outer Continental Shelf. Historic Preservation Fund, https://home.nps.gov/subjects/historicpreservation/historic-preservation-fund.htm, accessed August 10, 2022.

Appendix

Brief History of Energy Timeline Primarily in the United States

1100	Iron blast furnace operations using charcoal in China
c.1350	Iron blast furnaces in Europe use charcoal until 1600s, when forests begin to deplete
1641	John Winthrop builds the iron furnace at Saugus, Massachusetts
1690	Coal begins to replace wood as major source of heat and cooking throughout Europe, partly due to forest depletion
1712	Installation of "Newcomen" engine in a coal mine to pump water
1742	Benjamin Franklin designs a metal-lined "fireplace" to make home heating more efficient (known as the Franklin Stove)
1776	American Revolution begins
1776	Invention of Watt's steam engine, which improved upon the Newcomen design
1809	British chemist Humphrey Davy develops the electric arc lamp
1839	First photovoltaic cell invented by French physicist Alexandre Edmond Becquerel to create an electric current from light
1859	Oil discovery in Titusville, Pennsylvania
1870	Standard Oil founded by John D. Rockefeller and Henry Flagler
1876	World's Fair Centennial Exposition's Machinery Hall featuring steam power
1876	Menlo Park laboratory completed in New Jersey
1879–1880	Edison's lab develops a filament to perfect a long-lasting incandescent light bulb
1882	First hydroelectric power plant on the Fox River in Wisconsin; Edison Electric Light Company builds its first commercial system with a central generation system on Pearl Street in New York City

1885	Internal combustion engine patented by Germans Gottlieb Daimler and Wilhelm Maybach
1888	George Westinghouse purchases Nicola Tesla's patents to the AC motor
1893	World's Fair Columbian Exposition, Chicago, featuring electrical power in the "Great White City"
1895/6	Hydroelectric power plant at Niagara Falls with Westinghouse AC system
1901	Nikola Tesla patents first solar panel
1904	Prince Piero Ginori Conti invents the first geothermal power plant at the Larderello dry steam field in Tuscany Italy
1908	Ford's Model T released on the market
1911	Congress creates the United States Bureau of Reclamation, which will build dozens of hydroelectric dams across the United States over the next several decades
1917	United States enters World War I
1933	Congress passes the Tennessee Valley Authority Act, which will authorize a federally owned utility to development a hydroelectric in the South
1933	Construction of 100-kw windmill power station in Crimea, Russia
1935	Public Utility Holding Company Act (Wilbur-Rayburn Bill) to control electrical rates
1935	Establishment of the Rural Electrification Administration as part of President Franklin D. Roosevelt's New Deal
1939	German Otto Hahn discovers process of nuclear fission
1941	Construction of Smith Putnam Wind Turbine in Vermont, the largest wind turbine for decades
1941	Bombing of Pearl Harbor, Hawaii, and United States enters World War II
1942	President Roosevelt authorizes the Manhattan Project, a top-secret government program to develop the atomic bomb
1945	Atomic bomb detonated in New Mexico
1945	End of World War II; United States drops atomic bomb on Hiroshima and Nagasaki, Japan
1947	Atomic Energy Commission, a civilian agency, formed to oversee peacetime development of nuclear energy
1951	Nevada Test Site; nuclear power produced for electricity in Idaho
1954	Development of the first *silicon* photovoltaic (PV) cell at Bell Labs capable of converting enough solar energy to run appliances
1965	Northeast blackout
1969	Congress passes National Environmental Policy Act
1970	First Earth Day
1970	Congress passes Clean Air Act
1972	Congress passes Clean Water Act
1973	Yom Kippur War prompts the Organization of Petroleum Exporting Countries (OPEC) oil embargo and subsequent energy crisis
1975	NASA's wind turbine program begins
1977	New York City blackout

1978	Public Utility Regulatory Policies Act
1979	Nuclear accident at Three Mile Island
1980s	United States provides tax credits for renewable energies
1986	Nuclear accident at Chernobyl, Ukraine (then Soviet Union)
1993	First commercial wind turbine in New Zealand
1996	California legislature deregulates utilities
2000	Power shortages in California force rolling blackouts
2001	Energy company in California declares bankruptcy
2003	Blackout across northeastern United States and parts of Canada
2011	Nuclear accident at Fukushima, Japan, following earthquake and tsunami
2012	Residential solar systems become affordable
2015	Tesla Motor Company launches solar power storage battery for use in individual households
2016	End of Cape Wind project
2021	Electrical infrastructure fails in Texas following three winter storms, affecting five million people.

Even the most basic discussions about electricity might need some familiarity with terms with which most energy users are only tangentially familiar. The following is a short glossary of terms:

Current: The current is used to describe the flow of electrical charges.

Electrification: Electrification is the act of electrifying, the building of electrical infrastructure, and the distribution and use of electrical power.

Load: The load is the amount of current supplied by a source of electrical power and carried by an electrical system. The term also refers to the various factors that may cause stress on an electrical system or on a structure. These include electrical voltage, physical pull, wind pressure, climate, and usually most significantly, market demand. This may include the variations of use over a twenty-four-hour day. Utilities calculate load in kilowatts (kw), a unit of measurement that measures the power of an electrical current.

Kilowatt (kw): A kilowatt is a unit of electrical power equal to 1000 watts.

Kilowatt-hour: A kilowatt-hour is the unit of electrical energy generated or used equal to one kilowatt acting for one hour. A kilowatt-hour (kwh) measures the quantity of power consumed per hour and charged to each customer accordingly. For example, in 1938 lights and small appliances operated at about 30 kwh a month, a refrigerator was estimated at 50 kwh, a range at 150 kwh, and a water pump at 10–25 kwh.

Voltage: Voltage measures the force of a current. A volt is the unit of measurement.

Watt: A watt is a unit of electrical power. Today, an average table lamp uses anywhere from 60 to 75 watts, while a large appliance like a refrigerator may require 600–1000 watts.

Such appliance and household energy requirements have changed, as new technologies have required more energy.[1]

Note

1. Leah S. Glaser, *Electrifying the Rural American West: Stories of Power, People, and Place* (Lincoln: University of Nebraska Press, 2009), xi.

Recommended Reading

Annotated Bibliography of Resources on Energy and Energy Use

Energy History in American Society

I have placed an asterisk next to possible "gift shop" books. Several energy histories of places outside the United States have published books in the last few years. These include neighbors like Mexico. German Vergara, *Fueling Mexico: Energy and Environment 1850–1950*. Studies on Environment and History Series. Cambridge: Cambridge University Press, 2021; Diana Montano's *Electrifying Mexico: Technology and the Transformation of a Modern City*. Chicago: University of Chicago Press, 2021; and Myrna Santiago's *The Ecology of Oil: Environment, Labor, and the Mexican Revolution, 1900–1938*. Cambridge: Cambridge University Press, 2009.

General Energy/Technology/Policy

Black, Brian, and Donna Rilling, eds. "Energy in Pennsylvania History." *The Pennsylvania Magazine of History and Biography* (Special Issue), 89:3 (October 2015).

*Cowan, Ruth Schwartz, and Matthew Hersh. *A Social History of American Technology*. New York: Oxford University Press, 1996. Reviews 250 years of technology as part of American culture. Discusses systems, automobiles, aviation, communications, and even biotechnology in the context of intellectual and artistic movements across three time periods: colonial, industrial, twentieth century.

Heinberg, Richard. *The End of Growth: Adapting to Our Economic Reality*. Gabrioloa Island, BC, Canada: New Society Publishers, 2011. Heinberg makes a compelling argument about why, even with a transition away from fossil fuels, human societies will simply have less energy and will need to rethink economic growth as tied to energy use.

Hunter, Lewis C., and Lynwood Bryant. *A History of Industrial Power in the United States, vols. 1, 2, & 3*. Charlottesville: University of Virginia Press, 1979, 1991. Hunter synthesizes years of research in a series that serves as a standard overview of technology in the United States.

*Jones, Christopher. *Routes of Power: Energy and Modern America*. Cambridge, MA: Harvard University Press, 2014. Jones argues that infrastructure (roads, canals, pipelines, and transmission lines) "intensified" dependence on fossil fuels. He believes that new, efficient, renewable energy infrastructure will help transition away from fossil fuels.

Klein, Maury. *The Power Makers: Steam Electricity, and the Men Who Invented Modern America*. New York: Bloomsbury Press, 2008. Klein explores how science and technology transformed industry and business models.

*Lifset, Robert. *American Energy Policy in the 1970s*. University of Oklahoma Press, 2014. Edited volume divided into four parts: political leadership, foreign policy, supply, and demand.

*Melosi, Martin V. *Coping with Abundance: Energy and the Environment in Industrial America*. New York: Alfred A. Knopf, 1985. Melosi covers the history and social impact of various sources of energy including coal, oil, gas, electric power, nuclear power, and public power.

Melosi, Martin V. *Effluent America: Cities, Industry, Energy, and the Environment*. Pittsburgh: University of Pittsburgh Press, 2001. This anthology on pollution in the cities models local stories.

*Nye, David E. *Consuming Power: A Social History of American Energies*. Cambridge, MA: The MIT Press, 1998. Reviews society's use of energy as a product of cultural choices, not technological advancement. Book sections correspond to energy sources.

*Rhodes, Richard. *Energy: A Human History*. New York: Simon and Schuster, 2018. Rhodes provides a broad overview in a publicly accessible format.

*Webber, Michael E. *Power Trip: The Story of Energy*. Basic Books, 2019. In this highly accessible overview aimed at addressing climate change, Webber points out that transitioning to a single energy source does not address the problem. Energy systems are integrated. He emphasizes, too, that individual choices are not enough without global awareness.

Williams, James. *Energy and the Making of Modern California*. Akron, OH: University of Akron Press, 1997. Comprehensive account of the development of energy sources in the state of California, which, in many cases, served as a testing ground and model for energy systems all over the country.

General-Cultural

Hughes, Thomas Parke. *American Genesis: A Century of Invention and Technological Enthusiasm, 1870–1970*. New York: Viking University Press, 1989. Hughes's overview discussed how new devices reflected the processes of inventors, scientists, and engineers who expressed human values "to organize the world for problem solving so that goods and services can be invented, developed, produced, and used." The late nineteenth century's celebrated independent inventors of popular history provided the foundations for industry and future "system-builders."

Marx, Leo. *The Machine in the Garden: Technology and the Pastoral Ideal in America*. **Oxford University Press, 1964.** While it does not actually discuss machines, Marx's book serves as a classic work explaining an American aesthetic that attempts to reconcile a conundrum about how Americans embraced industry in what was supposed to be an agrarian nation.

Nye, David E. *American Technological Sublime*. **Cambridge, MA: The MIT Press, 1994.** Picking up on a phrase of earlier historians, Nye examines people's fascination with the aesthetics of technology and what it symbolizes about American values.

Wuebben, Daniel L. *Power Lined: Electricity, Landscape, and the American Mind*. **Lincoln: University of Nebraska Press, 2019.** Wuebben examines overhead wires as cultural symbols defining industrial spaces and shaping the modern American sensibility and mind.

Animal/Muscle

Greene, Ann Norton. *Horses at Work: Harnessing Power in Industrial America*. **Cambridge, MA: Harvard University Press, 2008.** Norton disputes the conventional narrative of industrialization—"machine replaces muscle"—by demonstrating that, contrary to popular and scholarly belief, the first wave of industrialization had the opposite effect on the use of animal energy.

Wood

Gordon, Robert. *A Landscape Transformed: The Ironmaking District of Salisbury, Connecticut*. **New York: Oxford University Press, 2000.** While this volume overlooks the larger context of energy use represented by such industrial sites, someone looking for energy stories can find abundant information.

Radkau, Joachim. *Wood: A History*. **New York: Wiley, 2012.** This is a sweeping, global cultural history from the "wood age" all the way to the present.

Native Americans and Energy

Allison III, James Robert. *Sovereignty for Survival: American Energy Development and Indian Self-Determination*. **New Haven: Yale University Press, 2015.** Allison convincingly argues that the Northern Cheyenne's and Crow's earlier and ongoing local efforts to control energy development, including renewable energy, on their reservation refined and expanded the definitions of Indian sovereignty, per the Indian Self-Determination Act.

Ambler, Marjane. *Breaking the Iron Bonds: Indian Control of Energy Development*. **Lawrence: University of Kansas Press, 1990.** Ambler reviews energy policies for American Indians and the legal history of Native control over natural energy resources, through the Council of Energy Resource tribes in the 1970s. Native groups began to assess, defend, and control energy development on their reservations.

Chamberlain, Kathleen. *Under Sacred Ground: A History of Navajo Oil, 1922–1982*. **Albuquerque: University of New Mexico Press, 2000.** This is an ethnohistory about how gas and oil production and leasing have impacted the Navajo Nation.

Duarte, Marisa Elena. *Network Sovereignty: Building the Internet Across Indian Country*, Seattle: University of Washington Press, 2017. Duarte argues that building internet infrastructure for Native communities reaches beyond issues of social and economic justice. ICTs (information/communication technology) can serve goals of cultural and political sovereignty.

Powell, Dana E. *Landscapes of Power: Politics of Energy in the Navajo Nation.* Durham, NC: Duke University Press, 2018. Powell examines the energy politics around the controversial Desert Rock power plant.

Smith, Sherry L., and Brian Frehner, eds. *Indians and Energy: Exploitation and Opportunity in the American Southwest.* Santa Fe, NM: School of Advanced Research Press, 2010. This collection of essays examines multiple aspects of energy development across reservations: development, use, impact of health and economy, etc.

Voyles, Traci Brynn. *Wastelanding: Legacies of Uranium Mining in Navajo Country*, Minneapolis: University of Minnesota Press, 2015. An "environmental justice" history that explores the political economy and equity issues around energy.

Fossil Fuels

*Adams, Sean Patrick. *Home Fires: How Americans Kept Warm in the Nineteenth Century.* Baltimore, MD. Johns Hopkins University Press, 2014. Adams provides an overview about how Americans developed coal as a primary source of heat energy for their homes. He does a fantastic job of putting this in a much wider historical context of energy, economic, and urban history. Adams also has several articles about consumerism and energy choice in the nineteenth century.

Adams, Sean Patrick. *Old Dominion Industrial Commonwealth: Coal, Politics, and the Economy in Antebellum America.* Baltimore, MD: Johns Hopkins University Press, 2009. Adams compares the political economies of coal in Virginia and Pennsylvania before the Civil War, and how the mining of this energy resource contributed to sectional differences.

Black, Brian. *Crude Reality: Petroleum in World History.* Baltimore, MD: Johns Hopkins University Press, 2003. Black looks further at the impact of petroleum from local (Pithole, PA) to global context.

Black, Brian. *Petrolia: The Landscape of America's First Oil Boom.* Lanham, MD: Rowman and Littlefield, 2003. Black explores the history of oil exploration and exploitation through the lens of place.

Blanchard, Charles. *The Extraction State: A History of Natural Gas in America.* Pittsburgh: University of Pittsburgh Press, 2021. Not authored by a historian or other scholar, but rather an industry expert, this work thus lacks scholarly rigor and documentation, but it is most comprehensive account of this topic to date.

Freese, Barbara. *Coal: A Human History.* New York: Basic Books, 2016. A popular history, this is a highly accessible overview that goes beyond the technical.

Frehner, Brian. *Finding Oil: The Nature of Petroleum Geology, 1859-1920.* Lincoln: University of Nebraska Press, 2011. Frehner explores the emerging science, the scientists, and prospectors behind oil extraction and how that influenced the industry.

Hunter, Lewis C. *Steamboats on the Western Rivers: An Economic and Technological History.* Cambridge, MA: Harvard University Press, 1949. Hunter describes the technological history of steamboats and their role as carriers of commerce across the expanding United States. He details every aspect of riverboating from engineering structure to passenger life.

Johnson, Bob. *Carbon Nation: Fossil Fuels and the Making of American Culture.* Lawrence: University of Kansas Press, 2014. Johnson dives deep into the history of fossil fuel use outside of technological history and opens not only cultural, but philosophical and physiological implications on society and our practices.

Lemenager, Stephanie. *Living Oil: Petroleum Culture in the American Century.* Oxford University Press, 2014. A humanities approach to exploring oil and its sensory, cultural, and aesthetic impact through art, literature, film, etc.

Malm, Andreas. *Fossil Capital: The Rise of Steam Power and the Roots of Global Warming.* London: Verso Books, 2016. This is the most comprehensive overview of fossil fuels that, with a nontechnical popular audience in mind, ties the history directly to climate change concerns.

Morris, Ian. *Foragers, Farmers, and Fossil Fuels.* Princeton, NJ: Princeton University Press, 2015. How energy systems reflect our values.

Pratt, Joseph A., Martin V. Melosi, and Kathleen Brosnan, eds. *Energy Capitals: Local Impact, Global Influence.* Pittsburgh: University of Pittsburgh Press, 2014. Intersection of fossil fuel development and urbanization, featuring specific cities, but likely apply to others.

Pursell, Carrol W. *Early Stationary Steam Engines in America: A Study in the Migration of Technology.* Washington, DC: Smithsonian Institution Press, 1962. Much of this study compares American technology with the English during the Industrial Revolution. However, Pursell's main argument is that, unlike in New England, the steam engine was vital in antebellum America for manufacturing in places where the water supply was unpredictable or undependable—specifically in the Old Northwest/Midwest.

Rose, Mark H. *Cities of Light and Heat: Domesticating Gas and Electricity in Urban America.* University Park: Penn State University Press, 1995. Rose reviews the history of municipal services in Denver and Kansas City—two booming Western cities relying on Eastern capital and innovation. He tries to address whether technology shapes society or society shapes technology, while mixing biography, technical data, gender issues. and urban development to illustrate the historical impact of domesticating energy. Also see Tobey and Cowan.

Rosen, William. *The Most Powerful Idea in the World: A Story of Steam, Industry, and Invention.* Chicago: University of Chicago Press, 2010.

Sabin, Paul. *Crude Politics: The California Oil Market, 1900–1940.* Berkeley: University of California Press, 2005. Sabin presents a study of oil dependence as a political process.

Sabin, Paul. "'The Ultimate Environmental Dilemma': Making a Place for Historians in the Climate Change and Energy Debates," *Environmental History,* Vol. 15, No. 1 (January 2010), 76–93.

Shulman, Peter. *Coal and Empire: The Birth of Energy Security in Industrial America.* Baltimore: Johns Hopkins University Press, 2015.

Nuclear Power

Allee, Glenna Cole, and Mark Auslander. *Hanford Reach: In the Atomic Field.* Durham, NC: Daylight Books, 2021. This book displays images of landscapes near Hanford, Washington.

Blaugh, Brian. *Chain Reaction: Expert Debate and Public Participation in American Commercial Nuclear Power, 1945–1974.* Cambridge: Cambridge University Press, 1991. This is one of the few works that addresses post World War II nuclear energy in a nonmilitary context.

Brown, Kate. *Plutopia: Nuclear Families, Atomic Cities, and the Great Soviet and American Plutonium Disasters.* Oxford: Oxford University Press, April 2013. Brown presents a transnational approach to the nuclear-arms race, but she focused on domestic threats to the environment and public health.

Cantelon, Philip, and Robert Williams, *Crisis Contained: The Department of Energy at Three Mile Island.* Carbondale: Southern Illinois University Press, September 1982. This account, using unpublished archival documents, concludes that the actual level of danger was not effectively communicated, and scientists and politicians responded in a way that alarmed the public so that it forever damaged support as an alternative energy source.

Cooke, Stephanie. *In Mortal Hands: A Cautionary History of the Nuclear Age.* New York: Bloomsbury, 2009. This is a comprehensive account by a journalist and industry insider.

*Findley, John, and Bruce Havly. *Atomic Frontier Days: Hanford and the American West.* Seattle: University of Washington Press, 2011. This work provides a local and community look at the impact of the atomic age at a site of production, combining the fields of science and technology with urban and sociocultural history.

*Fox, Sarah Alisabeth. *Downwind: A People's History of the Nuclear West.* Lincoln, NE: Bison Books, 2014. Based on interviews, this work gives voice to those impacted economically and regarding public health.

Gerber, Michele Stenehjem. *On the Home Front: The Cold War Legacy of the Hanford Nuclear Site.* Lincoln: University of Nebraska Press, 1992.

Hepner, Abbey, Kirsten Pai Buick, and Nancy Zastudil. *The Light at the End of History.* Durham, NC: Daylight Books, 2021. Hepner visually reveals the human and environmental relationships with nuclear technologies, through the very meaningful medium of uranium prints.

Kirk, Andrew G., and Kristian Purcell. *Doom Towns: The People and the Landscapes of Atomic Testing: A Graphic History.* Cambridge: Oxford University Press, 2017. This innovative model tells the story through drawings, but Kirk contextualizes it all with primary documents about those most affected by nuclear weapons during the Cold War.

Mahaffey, James. *Atomic Accidents: A History of Nuclear Meltdowns and Disasters from the Ozark Mountains to Fukushima.* New York: Pegasus Books, 2015. Similar to Stephanie Cooke, Mahaffey dives deep into the nature of nuclear accidents and how they have affected the technology.

Mahaffey, James. *Atomic Awakenings: A New Look at the History and Future of Nuclear Power.* New York: Pegasus Books, 2010. A research scientist provides a robust and comprehensive survey.

Melosi, Martin V. *Atomic Age America.* Boston: Pearson, 2012. Melosi provides a very useful and sweeping technological, political, and cultural overview addressing military and domestic issues of atomic energy from atomic theory to Fukishima.

*Pritikin, Trisha. *The Hanford Plaintiffs: Voices from the Fight for Atomic Justice.* Lawrence: University of Kansas, Press, 2020. Oral history documents the far less celebratory side of nuclear power development and the human cost of those closest to the sites of extraction, production, and use.

Rothman, Hal K. *On Rim and Ridges: The Los Alamos Area since 1880*. Lincoln: University of Nebraska Press, 1992. Although very much an environmental history, Rothman reviews an area where scientists conducted the most notorious scientific nuclear experiments and its effect on the landscape. He also notes that the residents and newcomers have battled for local resources. This book serves as a good companion to John Findlay and Bruce Havley's *Atomic Frontier Days*.

*Walker, J. Samuel. *Three Mile Island: A Nuclear Crisis in Historical Perspective*. Berkeley: University of California, 2004. Former US Nuclear Regulatory Commission historian used NRC and records from the President's Commission on the accident as well as presidential papers to explain that "the accident at Three Mile Island (TMI) in March 1979 embarrassed promoters of commercial nuclear energy, reconfirmed the protests of antinukes, and set back the once revolutionary power source—in the U.S. at least—for decades."

Wellerstein, Alex. *Restricted Data: The History of Nuclear Secrecy in the United States*. Chicago: University of Chicago Press, 2021. Born from blog of the same name that posted and discussed recently declassified documents, Wesserstein, a consultant with the Atomic Heritage Foundation, writes about the secrecy around the atomic bomb in terms of its development and effects from the 1930s through the end of the Cold War.

Wellock, Thomas R. *Critical Masses: Opposition to Nuclear Power in California, 1958–1978*. Madison: University of Wisconsin Press, 1998.

Zoellner, Tom. *Uranium: War, Energy, and the Rock that Shaped the World*. New York: Penguin, 2009.

Electricity/Electrification

Bowers, Brian. *Lengthening the Day: A History of Lighting Technology*. Oxford University Press, 1998.

Brigham, Jay L. *Politics and Power: The Fight for Electricity in the West, 1902–1932*. Lawrence: University of Kansas Press, 1998. Brigham argues that public power was a vital political issue prior to the Great Depression and that New Deal reforms were merely the manifestations of arguments made during the progressive era.

Brown, D. Clayton. *Electricity for Rural America: The Fight for the REA*. Westport, CT, London England: Greenwood Press, 1980. Technological historians consider Brown's book a standard for the early development and activities of the Rural Electrification Administration.

Casazza. *Development of Electrical Power Transmission: The Role Played by Technology, Institutions, and People*. Piscataway, NJ: IEEE Press, 1993. This trade publication that provides an excellent timeline.

Cater, Casey P. *Regenerating Dixie: Electric Energy and the Modern South*. Pittsburgh: University of Pittsburgh Press, 2019. Cater fills a regional gap in the growing historiography about electricity and the electrical industry in America by reframing the popular understanding about the modernization and industrialization of the "New South."

Childs, Marquis. *The Farmer Takes a Hand: The Electric Power Revolution in Rural America*. New York: Doubleday, 1952. This is one of the few works on rural electrification that concentrates on social and political issues rather than institutional and administrative ones.

Cooper, Gail. *Air-Conditioning America: Engineers and the Controlled Environment, 1900–1960*. Baltimore, MD: Johns Hopkins University Press, 1998. Interesting example of how the marketing of technology shapes daily life and values. Developed early for humidity control in industrial spaces, indoor climate control was slowly accepted by the public, as Cooper reviews. New building types after World War II included central air units, along with cheaper window units, and helped popularize the notion.

Freeburg, Ernest. *The Age of Edison: Electric Light and the Invention of Modern America*. New York: Penguin, 2014. This is a popular history and synthesis that places lighting within social historical context. Provides a good introduction and primer for those new to the topic.

French, Daniel. *When They Hid Fire: A History of Electricity and Invisible Energy*. Pittsburgh: University of Pittsburgh Press, 2017. French combines cultural and technological to make the argument that, "As electricity became disassociated with coal in the minds of Americans, an ideology of energy exceptionalism reformed around renewed beliefs of inconsequential consumption."

Friedel, Robert, and Paul Israel, *Edison's Electric Light: Biography of an Invention*. New Brunswick, NJ: Rutgers University Press, 1986 (updated as *Electric Light: The Art of Invention*. Baltimore, MD: Johns Hopkins University Press, 2010). This is a study about the process of invention as much as the story of the technology. Friedel and Israel primarily studies Edison's papers at Menlo Park and the e-book version connects directly to digital versions of the primary sources they used.

Glaser, Leah S. *Electrifying the Rural American West: Stories of Power, People, and Place*. Lincoln: University of Nebraska Press, 2009. Glaser provides a social and cultural history of rural electrification in the West. Using three case studies in Arizona, Leah S. Glaser details how, when examined from the local level, the electrification process illustrates the impact of technology on places, economies, and lifestyles in the diverse communities and landscapes of the American West.

Hirt, Paul. *The Wired West: The History of Electric Power, 1870s–1970s*. Lawrence: University Press of Kansas, 2012. Hirt examines the one hundred years from the origins of electricity in the Pacific Northwest—defined as Oregon, Washington, Idaho, and British Columbia, Canada. Hirt uses a comparative approach and transborder framework enabling a study about the importance of place in the development of electrical power within national contexts. He identifies geography, natural resources, and culture as factors that shaped how electrical power systems developed.

Hoy, Suellen M., Michael C. Robinson, and Ellis L. Armstrong. *History of Public Works in the United States, 1776–1976*. Chicago: American Public Works Association, 1976. A trio of historians, hired by the American Public Works Association, have written a complete and comprehensive overview of public works in the United States.

Hughes, Thomas. *Networks of Power: Electrification in Western Society, 1880–1930*. Baltimore and London: The Johns Hopkins University Press, 1983, 1988. This is the definitive, though very thick, work on the history of electricity. With both engineering and historical training, Hughes places electrical technology within the context of politics, commerce, and society.

Keating, Ann Durkin. *Invisible Networks: Exploring the History of Local Utilities and Public Works*. Malabar, FL: Krieger Publishing Company, 1994. Keating reviews how public works structures can be used to research community histories. She highlights the historical importance of many of these engineering structures and the need to preserve them. This book provides a guide for beginning researchers in this field.

Melosi, Martin. *Thomas A Edison and the Modernization of America*. Glenview, IL: Scott Foresman, 1990. A concise synthesis biography within the contest of industrial and urban history.

Myers, William A. *Iron Men and Copper Wires: A Centennial History of the Southern Edison Company*. Glendale, CA: Trans-Anglo Books, 1983. Myers provides a full and detailed account that describes the evolution of the Southern Edison Company in California and provides a guide for other technological and institutional histories. Places the story within the context of the evolution of electrical power.

Needham, Andrew. *Power Lines: Phoenix and the Making of the Modern Southwest*. Princeton, NJ: Princeton University Press, 2014. Needham reframes postwar urban development in the American West and even sheds some light (literally and metaphorically) upon the rural-urban divide that our current political climate suggests. The growth of cities like Phoenix were a phenomenon profoundly affecting urban residents, but in many ways the rural, largely Native American, populations and their environments even more acutely as utilities and the federal government built a vast regional power network.

Nye, David E. *Electrifying America: Social Meanings of a New Technology, 1880–1940*. Cambridge, MA: The MIT Press, 1990. Nye's research and interpretations are indispensable work on the social and technological history chronicling the effect of electricity on America. He concentrates primarily in the Midwestern and Western regions. Nye's primary questions focus on what people use electricity for, and how it affected social human patterns and landscapes.

Nye, David E. *When the Lights Went Out: A History of Blackouts in America*. Cambridge, MA: MIT Press, 2010. Beginning with a concise history of the grid, Nye reviews the various times we have lost light, intentionally for issues of security, or accidentally, often leading to serious crises, how people responded, and why.

Spence, Clark C. "Early Uses of Electricity in American Agriculture," *Technology and Culture* 3, Spring 1962. Spence fills this article with little known information of how farms used electrical power prior to full-scale electrification.

Tobey, Ronald. *Technology as Freedom: The New Deal and the Electrical Modernization of the American Home*. Berkeley: University of California Press, 1995. Tobey discusses how the concepts of modernization introduced during the New Deal changed postwar American home life aided by new housing programs. Extensive statistical data and interesting discussions about how policy and other issues like gender, class, and race affected decisions about domestic use of emerging electrical technology.

Zallen, Jeremy. *American Lucifers: The Dark History of Artificial Light, 1750–1865*. Chapel Hill: University of North Carolina Press, 2019. Zallen shifts the emphasis from inventors and industrialists to examine this topic through the lens of labor and inequality, addressing the sacrifices of free and unfree labor, such as whalers, child industrial workers, seamstresses, and coal miners

Energy Industry and Business

Carlson, W. Bernard. *Tesla: Inventor of the Electric Age*. Princeton, NJ: Princeton University Press, 2013. Contextualizing the inventor, Carlson examines the idea of invention as both a scientific and economic process.

Funigiello, Philip. *Toward a National Power Policy: The New Deal and the Electric Utility Industry, 1933–1941*. Pittsburgh: University of Pittsburgh Press, 1973.

Hirsh, Richard F. *Technology and Transformation in the American Electric Utility Industry*. Cambridge, MA: MIT Press, 1989.

Jonnes, Jill. *Empires of Light: Edison, Tesla, Westinghouse, and the Race to Electrify the World.* New York: Random House, 2004. Popular and very accessible history of the competitive war between DC current and AC current.

Komarek De Luna, Phyllis. *Public Versus Private Power During the Truman Administration: A Study of Fair Deal Liberalism.* Modern American History Series. New York and Bern: Peter Lang, 1997. Komarek describes the political struggle over public power, begun in the 1920s and 1930s, over who, mainly, should generate electricity, and she uses the controversy to test the extent of President Harry Truman's "Fair Deal" liberalism in its role as legatee of Franklin Roosevelt's New Deal.

Neufeld, John L. *Selling Power: Economics, Policy, and Electric Utilities Before 1940.* Chicago: University of Chicago Press, 2016.

Energy and Gender

Cowan, Ruth Schwartz. *More Work for Mother: The Ironies of Household Technology from the Open Hearth to the Microwave.* New York: Basic Books, 1985. Revealing work on the relationship between gender and technology, Schwartz argues that household appliances altered the roles and relationships in the family and home and shifted most tasks to the woman who presumably had various devices to aid her. Suburbanization helped spur the industrialization of the home. Devices used in the privacy of one's home grew more popular than similar services that people outside the offered like laundry and cooking. New technologies also raised expectations.

Jellison, Katherine. *Entitled to Power: Farm Women and Technology, 1913–1963.* Chapel Hill: University of North Carolina Press, 1993. This very readable monograph describes how the advent of modern agribusiness changed patterns of life and labor on the family farm. As women adopted laborsaving devices, they were able to spend more time working independently on the farm itself or within the community rather than within their prescribed domestic sphere.

Water and Hydropower

Jackson, Donald C. *Building the Ultimate Dam: John S. Eastwood and the Control of Water in the West.* Lawrence: University Press of Kansas, 1995. Jackson discusses the construction of low-cost multiarched dams for private corporations and the interrelationship between technology, capitalism, and politics in building such massive structures through the career of innovative engineer John S. Eastwood.

Linenberger, Toni Rae, and Leah S. Glaser, *Dams, Dynamos, and Development: The Bureau of Reclamation's Power Program and Electrification of the West.* Washington, DC: US Government Printing Office, 2002.

Manganiello, Christopher J. *Southern Water, Southern Power: How the Politics of Cheap Energy and Water Scarcity Shaped a Region.* Chapel Hill: University of North Carolina Press, 2015. A companion to Cater's *Regenerating Dixie*, Manganiello explores and explains how aggressive use and management of water resources caused a non-arid region came to have issues with water scarcity.

Pitzer, Paul C. *Grand Coulee: Harnessing a Dream*. **Pullman: Washington State University Press, 1994.** This is a celebratory and detailed account of the building of one of the largest hydroelectric dams.

Steinberg, Theodore. *Nature Incorporated: Industrialization and the Waters of New England.* **Amherst: University of Massachusetts Press, 1991.** Steinberg reviews control of the Merrimack River for industrial use, economic growth, with attention to the environmental impact on the region.

White, Richard. *The Organic Machine*. **New York: Hill and Wang, 1995.** White recounts how mechanical control over the Columbia River depleted the once abundant salmon population. He primarily concentrates on what he considers the natural relationship between humans and the environment and how humans have utilized the energy of the river. White contends that this instance is only one example of how changing perceptions about nature, technology, and progress over time can have dire effects on nature and the ecosystem in which humans live.

Wind and Renewables

Baker, T. Lindsay. *A Field Guide to Windmills*. **Norman: The University of Oklahoma Press, 1985.** Baker provides a comprehensive overview of the history of windmill use, design, and manufacturing, and marketing across the United States. This is a definitive work on the subject as both a reference and a model for researching and writing technological history from material culture.

Barber, Daniel A. *A House in the Sun: Modern Architecture and Solar Energy in the Cold War.* **Oxford University Press, 2016.**

Perlin, John. *Let it Shine: The 6,000-Year Story of Solar Energy*. **Novato, CA: New World Library, 2013.** While primarily a history of technology, rather than a historical narrative or context, it provides a timeline and with all of the examples of technology, plenty of ways to understand the material culture of solar.

Righter, Robert W. *Wind Energy in America: A History*. **Norman: University of Oklahoma Press, 1996.** This work reviews the history of wind power, focusing particularly upon the evolution from windmills as water pumpers to electrical generators. Righter accessibly provides political and economic historical context for this little studied energy source, observing that a commitment to the economy of scale and control offered by fossil fuels and centralized power grid systems.

Stove Bibliography

Groft, Tammis K. *Cast with Style: Nineteenth Century Cast-Iron Stoves from the Albany Area.* **Albany: SUNY Press, 1984.**

Harris, Howell John. "The Stove Trade Needs Change Continually": Designing the First Mass-Market Consumer Durable, ca. 1810–1930. *Winterthur Portfolio* 43:4 (2009), 365–406.

Harris, Howell John (2008). Conquering Winter: U.S. Consumers and the Cast-Iron Stove. *Building Research and Information* 36(4): 337–50.

Peirce, J. *Fire on the Hearth: The Evolution and Romance of the Heating-Stove*. **Springfield, MA: The Pond-Ekberg Co., 1951.**

Guides and Technical Bulletins

Moore, Jackson W., Jr. "Historical Utility Lines: A Resource in Search of Constituents," *CRM Bulletin* 3:2 (June 1980), 1, http://npshistory.com/newsletters/crm/bulletin-v3n2.pdf.

Sabin, Paul et al. "Energy History Online," https://energyhistory.yale.edu/. New Haven, CT: Yale University, accessed August 22, 2021. Provides various curated teaching units and materials surrounding energy history.

Whatley, Michael. *Interpretative Solutions: Harnessing the Power on Interpretation to Help Resolve Critical Resource Issues* (ebook). Natural Resource Stewardship and Science Office of Education and Outreach, and National Association for Interpretation, 2011.

Index

Atomic Testing Museum (ATM), Las Vegas, Nevada, 109, 120
automobiles, 62, 82, 92, 96, 97–98, 179; electric, 97

B Reactor, Hanford, WA, 110, 111–13
Baltimore Museum of Industry, 2
Barber, Daniel, 144
Beates, Susan, 93, 94
Beck, Larry, 5
Beckley Iron Furnace, 17, 18, 25
Bedford County, Pennsylvania, 138
bellows, 19, 38
Berks County, 19
Bethlehem Steel Company, 22
Bethlehem, Pennsylvania, 14
Bierstadt, 46
biofuel, 28, 98, 129, 132, 145
Black, Brian, 83, 91, 179
Blackberry River, 18, 25; blackouts, 157. *See also* Northeast Blackout of 2003
blacksmith, 13, 26, 93
Blanchard, Charles, 83, 182
bloomeries, 13
Boardman, Samantha, 95–96
Boott Museum and Boarding House, 41. *See also* Lowell Historical National Park
Boston, Massachusetts, 12, 17, 44
Boulton, Matthew, 81
Bowers, Brian, 84, 100
Bradshaw, Charles, 66
Brigham, Jay, 152, 185
briquettes, 25
Brochu, Lisa, 5
Brooke, James, 65
Brown, William, 39
Brownfields, charcoal, 11, 13, 19, 20. *See also* CERCLA
brownouts (2000), 157
burning for fuel, 1, 2, 3, 7, 11–13, 19, 22, 23, 25, 27, 52, 56, 57, 59, 66, 79, 81, 85, 129, 150, 172. *See also* firewood
Bush, George W., xiv

Cable, Ted, 5
Cabot Oil and Gas, 95
Cameron, Catherine, 68
canals, 41, 43–46
Cannon, Brian, 16
Cape Wind, Massachusetts, 133, 135, 136, 177
carbon dioxide, xiii, 11, 13, 79, 130, 145
carbon footprint (calculating), 4, 66
carbon, 66, 79, 80, 81, 82, 97, 105, 121, 130, 140. *See also* carbon emissions
cars. *See* automobiles
Carter, (President) James (Jimmy), 21
Casco Bay Line, 55
cast house, 13, 21
cast-iron, 12, 26, 27, 28, 164, 189; skillets/pots/pans, 13; *See also* stoves
Cater, Casey, 153, 185
Chamberlain, Kathleen, 82, 181
charcoal, 7, 11, 13–14, 16, 17, 19, 20, 23, 25–26, 27, 35, 49, 79, 80, 87, 150, 172, 175; making of, 13, 14, 17, 20, 21–23; *See also* charcoal-making
Charles Morgan (Ship), 61
Chaude-Aigues, France, 132
Chernobyl, Ukraine, 103, 177
Chicago, Illinois, 14, 24, 140, 151, 152, 176
China Syndrome (The), 106
China, xiii, 175
Church Rock, New Mexico, 107
Civilian Conservation Corps (CCC), 21
Clean Air Act (1970), 55, 66
clean energy (also see "green" energy), 34, 39, 56, 57, 79, 109, 133–38, 145
Clean Water Act (1972), 55, 56, 66, 176
climate change, xiii, xiv–xvii, 1–6
Clooney, George, 155
Coal and Coke Heritage Center, 86
coal-fire steam, 14, 27, 51, 53–54, 58, 65–67, 71–72
coal, 6, 7, 12, 13, 14, 25, 26–27, 34, 47, 51–72, 79–86; anthracite, 12, 14, 21, 25, 27, 79, 80, 81, 86, 95–96;

Dupont, Henry, 53–54
Durango and Silverton Narrow Gauge
    Railroad, 53, 61–67, *64*, *67*, 72
Durango Herald, 62
Durango, Colorado, 61–67, 69

Earth Day, xv, 55, 176
East Rock (Park), 36, 38, 40
Eastern Mining Area (EMA) Transmission
    Line, Arizona, 157–59
Eckley Miners' Village Museum, 86
economy of scale, 81, 131, 133, 149, 152
ecosystem, xiv, 34, 133, 135, 137, 189
Edison, Thomas, 40, 150–52, 160, 175
education, 4, 18, 35, 38, 40, 41, 43, 44, 53, 54,
    56, 57, 62, 68, 69, 84, 86, 87, 88–89, 92, 93,
    94, 121, 138, 139, 144, 156, 172;
  environmental/energy education, 4, 86, 93;
  STEM (Science, Technology, Engineering,
    Math), 39, 41, 87, 88, 116, 123, 162;
  STEAM, 39.
  See also science museums
electric motor/engine, 38, 40, 59, 97
electrical load, 162
electricity, xv, 3, 6, 7, 34, 39, 44, 47, 70, 79, 81,
    95, 99, 104, 105, 129–30, 140, 142, 149–66,
    177–79
Eli Whitney Museum and Workshop,
    Hamden, Connecticut, 36–41
emissions (carbon), xiv, 1, 14, 56, 57, 65–66,
    67, 69–72, 97, 121, 129, 136–37, 145, 171
Eneas, Aubrey, 132
energy choice, xiii-xv, 3–4, 6–8, 15, 23, 40–41,
    71, 80, 84, 86, 96, 97, 98, 99, 105, 110, 133,
    143, 145, 155, 180
energy efficiency, 1, 4, 14, 18, 26, 35, 69,
    70–71, 72, 162
energy transition, xv, 2, 3, 4, 12–13
energy, types of:
  chemical, 7, 11, 16, 118, 130, 145;
  gravitational, 7, 39, 130;
  kinetic, 4, 7, 28, 33–34, 36, 41–42, 51, 80,
    129–31;

mechanical, 7, 34–35, 39, 42, 51, 54, 81, 140,
    151, 152;
  motion, 7, 39, 53, 80;
  potential, 4, 7, 42;
  radiant (light), 7, 11;
  sound, 7, 19, 39, 53, 132, 137;
  thermal (heat), 7, 11–12, 26, 51, 79, 81,
    129, 132;
  See also wood, water, steam, fossil fuels,
    nuclear, renewable, electric
engine, 7, 132, 140;
  (coal-fired) steam engine, 11, 14, 19, 47,
    51–77, *54*, *59*, *60*, 81, 85, 96, 97, 150, 152,
    172, 175;
  combustion, 19, 61, 66, 96, 97, 176;
  diesel, 55–57, *59*, 62, 66, 67, 96
engineering, xiii, 41, 52, 57, 110, 116, 155
engineeriums, 53
Enola Gay (exhibit at National Air and Space
    Museum, Smithsonian Institution), 107–9,
    108, 123
environmental justice, xiv–xv, 105, 113, 121,
    135, 172, 182
Environmental Protection Agency (EPA), 66,
    109, 171
environmental resilience, 3, 15, 17–19,
    112, 117
environmental sustainability, xv–xvi, 1, 2, 26,
    58, 61, 68, 71, 91, 135, 172
Eric Sloane Museum, 19
Ericsson, John, 132
Ethanol, 34, 84, 132

Fads and Failures (exhibit), *97*
farmers, 11, 12, 15, 19, 43, 82, 135, 157
Federal Energy Regulatory Commission
    (FERC), 48
fireplace, 143, 164, 175. *See also* hearth
firewood, 12, 163, 164
Fitch, John, 52
Fleischer, Ari, xiv
flux, 13, 16, 25

food systems, production, and energy, 7, 40, 90, 130, 145, 149, 172;
  food preservation, 143, 156, 162, 164
forest, 1–25, 65, 66, 91, 94, 96, 145, 175;
  conservation, 17;
  deforestation, 12, 94, 96.
  See also trees and wood
Forest County Visitor Center, 94
forge (for iron tool production), 5, 6, 13, 15, 20, 26, 34, 38
fossil fuels, xiv, 2, 3, 6, 12, 21, 28, 51, 52, 67;
  histories of, 80–85;
  interpreting, 79–80, 85–96, 103;
  oil/petroleum, 6, 21, 34, 51, 53, 55, 67.
  See also coal, gas
Four Corners:
  tourism, 62;
  energy, 69;
  See also Colorado Plateau
Fox River Valley, 140, 175
Fox, Sarah A., 106
Fracking. *See* gas
Francaviglia, Richard, 15
Franklin, Benjamin, 12, 175–76;
  stoves, 26
Franklin, Robert, 118
Freese, Barbara, 82, 85
French Creek State Park, 20, 23
French, 132, 135, 175
French, Daniel, 84
Friedman, Thomas, 137
Friends of Beckley Furnace, 18
furnace:
  for iron production, xii, 2, 11, 12, 13–23, *17*, *18*, 26–28, 33–34, 56, 80, 175;
  home mechanical, 98, 99, 144

gas, xiv, 6, 7, 18, 34, 35, 51, 57, 62, 70, 79–80, 82, 83, 84, 87, 92–94;
  fracking, 91, 94, 95, 96;
  natural gas, 40, 66, 79, 80, 81, 86, 91, 94–96
Gasiorek, Chris, 59
Gatewood, John, 68

General Electric, 152. *See also* Thomas Edison
generator, 34, 80;
  diesel, 59;
  hydroelectric, 151, 160;
  wind, 133
George Wright Society, 1
Geothermal. *See* renewable energy
Gettysburg, Pennsylvania, 20
Gieringer, Laurence T., 95–96
Gilbert, A.C., 39, 40
Gilded Age, 149, 151
glassware, *119*;
  uranium glass, 120–21;
  See also Vaseline glass
Goodheart, Andrew, 90
Gordon, Robert, 14–15
Gore, Al (Vice President), 23
Grand Canyon, 33
green museum, xvii, 1
greenhouse gas, 57, 66, 143, 145
Gulf Coast, 80
Gusterson, Hugh, 109

Halladay, Daniel, 140
Ham, Sam, 5
Hamblin, Jacob Darwin, 107
Hanford Mills Museum, 1
Hanford, Washington, 110, 113, 117;
  collections, 118
Harper, Allen, 26, 66–67, 68
Harper's Weekly, 106
Harries, Mags, 44
Harrington, Lilly, 164
Harris, Howell, 26–27
Harrison, Leah G., 43–46
Harvey, Fred, 62
Hawkins, Harold, 20
Hearth, 12–13, 20, 21, 26–27, 80, 84, 98. *See also* stove and heating
heat, 7, 11. *See also* energy, thermal
heating, 7, 11, 51, 81, 84;
  of iron furnace, 13, 14, 16, 25;
  of fossil fuel, 7, 11, 79;

geothermal, 132;
  for home, 11, 12, 13, 23, 26–28, 40, 81, 84,
    99, 142–44, 162, 175;
  solar, 132, 144;
  of water, 34, 51
Héder, Lajos, 44
Heinberg, Richard, 80
Helen Thatcher White Foundation, 62
heritage railroads;
  Cumbres and Toltec Narrow Gauge
    Railroad/Scenic Railroad, 63, 66;
  Durango and Silverton, Narrow Gauge
    Railroad (D&SNGRR), 61–69;
  image in *New York Times*, 64, 65;
  pollution, 65.
  See also emissions, smoke
heritage tourism, 2, 88, 91, 136
Hills, Richard L., 52
Hiroshima, Japan, 105, 109, 112, 113, 116, 176
Hiroshima Peace Museum, Japan, 110
Hirt, Paul, 153
Historic American Engineering Record
  (HAER), 48, 72, 158
historic homes, 71, 72, 142–45
historic preservation, 2, 52, 71, 172;
  and solar energy, 138–40;
  in harmony with environmental initiatives,
    71–72, 142–44;
  in conflict with environmental initiatives, 2,
    56, 72, 135;
  architectural features, 142–45.
  See also Preservation Connecticut
Holley, Alexander Lyman, 15, 29
Home and Flower Exposition (1976), 144
Hope and Trauma in A Poisoned Land, 121
Hopewell Big Woods, 21
Hopewell Furnace National Historic Site,
  Ellsworth, Pennsylvania, 19;
  charcoal-making demonstration at, *22*;
  Establishment Day, 22;
  history of, 19–20;
  interpreting energy, 21–23;
  products, *24;*
  and the rural-industrial landscape, 20

Hopewell Lake, 20
Horseshoe Shoal, 133
Houck, Lafayette, 21
Hudson River Valley School, 46
Hughes, Thomas Parke, 152
Hunner, John, 110
Hunter, Louis C., 52
Hurt, R. Douglas, 52
hydraulic:
  fracturing, 34, 94, 95;
  turbine, 13, 38, 41
hydroelectricity, 34, 35–36, 41, 44, 46, 142,
  151, 154, 158–59, 160, 175–76;
  transmission line, 34, 40, 80, 82, 142, 151,
    152, 157–59;
  transmission towers, 159–60
Hydropower, 34, 35, 44, 46, 129, 140, 153;
  and hydroelectricity, 34.
  See also waterwheel and turbine

ice box, *161*
Iceland, 132
Indian Arts and Crafts Board, 166
Indian Mineral Leasing Act (1938), 83
industrial history, xv, 15, 20. *See also* Industrial
  Revolution
Industrial Revolution:
  in Britain/Europe, 51, 81;
  in United States, 46, 52, 149
Industry (extractive), 16, 85
infrastructure, 6, 80, 82–83, 133, 145, 149–57;
  household, 98–99
Insull, Samuel, 152
International Coalition of Sites of Conscience,
  6, 113
*Interpretation: Making a Difference on
  Purpose*, 5
Interpreting Our Heritage, 4–5
interpretive models, 4–5
Iran, 104
iron (appliance), 154, 162, 163–64
iron and iron production, 175;
  history, 13–19;
  interpreting, 15–27;

ore, 13;
pig, 13, 26;
production process, 13;
windmills, 140;
wrought, 13.
See also iron furnaces
iron furnaces, 2, 11–28, 33, 46, 53, 54, 80,
    91, 175;
Beckley, *17, 18*, 25, 26;
blast, 19;
blowers, 13, 19;
charging at, 13;
Hopewell, 17, 19–23;
ironmaster, 13, 16, 19;
operations, 13–28

Jacobs Wind Electric Company, 140
Jefferson, Thomas, 15;
    agrarian dream, 15
Jellison, Katherine, 156
Jensen, Oliver, 56, 62, 63
Johnson, Bob, 82, 90
Johnston, Barbara Rose, 106
Jones, Christopher, 82, 84
Jones, Kathryn, 64, 82, 85
Jonnes, Jill, 151

Keating, Ann Durkin, 35, 158
Kee Kilowatt, 154
Kelly, Cynthia, 109, 110. *See also* Atomic
    Heritage Foundation
Kent Iron Furnace, 19
Kentucky Coal Mining Museum, 86;
    and solar, 86
Kern River, 160
kerosene (fueled appliances), 84, 98, 164
kitsch, 88, 98, 118
Klamath Basin, Oregon, *131*
Klein, Maury, 52
Kline, Robert, 156
Kohl, Elmer, 22
Krupar, Jason, 113
Kyoto Protocol, 137

La Brea Tar Pits, Los Angeles, California, 90
la Cour, Poul, 140
labor, 7, 12, 14, 34, 84;
    extractive landscapes, 15, 19, 22, 27–28;
    and fossil fuels, 81, 85–86;
    and manufacturing, 36, 40, 41
lamps/lighting, 7, 26, 156, 158;
    histories of, 84, 87, 96, 98;
    lamps (fuel and labor), 84, 98–99, 154,
        162–64
landscape:
    agricultural, 90, 135–36;
    as artifact, 46–47;
    nuclear, 117–18;
    rural-industrial, 15;
    working, 15, 16, 19, 33, 47–48
Legacy Management, U.S. Department of
    Energy (Office of), 113–16, 117;
    Atomic Legacy Cabin, Grand Junction,
        Colorado, 113–16, *114, 116;*
    Fernald Preserve Interpretive Center,
        114–17, 126;
    Weldon Spring Interpretive Center, 115–17
LeMenager, Stephanie, 90
Lifset, Robert, 106
light, 2, 3, 7, 39, 79, 152, 153, 155, 164, 166;
    blackouts, 157;
    in houses, 142–44;
    solar, 130, 133;
    See also lightening
lightening, 166
limestone, 13, 20, 21, 25, 135, 140;
    as flux, 13
Linenthal, Edward, 83
Long, Harker, 21
Los Alamos History Museum, 113
Los Alamos National Laboratory, 109
Lovell, Tom, 90
Lowell National Historical Park, Lowell,
    Massachusetts, 2, 34, 41–43, 56;
    Suffolk Mills, 41–42, *42*
Lufkin, Daniel, 56

Madsen, Gary, 106

Magoc, Chris, 91–94

Mahaffey, James, 106

Malm, Andreas, 52

Malone, Patrick, 34

Manhattan Project (The), 104–6, 107, 108, 110, 112, 114–16

Manhattan Project National Historical Park, 109

Mann, Melissa, 93–94

Manuel, Jeffrey, 132

Marcellus Shale, 94–96

Mark, Leo, 15, 46, 71

material culture, 6, 10;
  analysis of, 7;
  Schlereth, Thomas, 8

Material Cultures of Energy, 7, 10

McFadden, Dan, 58

McKinney, Alexis, 74

Melosi, Martin, 106, 152

Melville, Herman (*Moby Dick*), 87

Menlo Park (Edison), 40, 150, 175

Merrill, Karen R., 83

Merrimack River, 41, 55

Merriman, Tim, 5

Mesa Verde, Colorado, 62, 71

Mill River, 36, 38

Mobilette Roadster, 97

Molly Maguires (The), 86

Moniz, Ernest, 93–94

Montana, 14

Montoya, Esperanza, 162

Moon, Michell, 3, 90, 171

Moran, Thomas, 46

Morris, Ian, 82

Mount Vernon, Virginia, 2

munitions, 16

Munson, William Giles, 46, *47*

Museum of History and Science, Seattle, Washington, 2

Museums of Western Colorado, 115

Mystic River, 54, 55

Mystic Seaport Museum, 53–61, *58, 59, 60, 61*, 71, 72. *See also* Sabino

Mystic, Connecticut, 53

Nagasaki Hanford Bridge Project Program, 113

Nagasaki, Japan, 105, 109, 110, 112, 113, 116, 118

Nantucket Sound, 133. *See also* Horseshoe Shoal

National Association for the Study and Prevention of Tuberculosis, 144

National Association of Interpretation (NAI), 1, 4–5, 144

National Atomic Museum, 109. *See also* National Museum of Nuclear Science and History

National Coal Association, 85

National Environmental Policy Act (1969), 55, 135, 176

National Historic Preservation Act, 48, 71, 72, 84, 158

National Museum of American History, 107, 155

National Park Service (NPS), 17, 20, 21, 35, 41, 62, 71, 110, 112, 171;
  and climate change, 1;
  and interpretative practices, 4, 171;
  and preservation policy, 72, 140.
  See also individual sites

National Science Foundation, 40

national security, xiv, xvi, 92, 103, 105, 137

National Trust for Historic Preservation, 1;
  "The Greenest Building is the One Already Built," 2

natural gas. *See* gas, fossil fuels

Navajo Generating Station (NGS), Page, Arizona, 154

Navajo Nation, 116, 121, 166. *See also* Diné

Navajo Tribal Utility authority (NTUA), 154

Needham, Andrew, 152–53

Nevada Test Site Historical Foundation (NTSHF), 109

Nevada Test Site, 103, 109, 117, 176

New Bedford Whaling Museum, 87

New Building Institute, 1

Post Carbon Institute, 80. *See also* Heinberg, Richard

Powell, Dana, 82

power grid, 131, 133, 142, 149, 151–57, 160

power lines, 2, 47, 133, 137, 152, 153, 166

Preservation Connecticut, 42, 138–40. *See also* historic preservation

Preservation Pennsylvania, 138. *See also* historic preservation

Priest, Tyler, 83

Pritikin, Trisha, 112–13

product catalog. *See* Sears, Roebuck, and Company

production (of energy), xiv, xv, 3, 6, 7, 13, 15, 19, 34, 38, 39, 41, 42, 44–45, 46, 48, 56, 80–85, 87, 91, 103–4, 111, 113–15, 121, 129, 133, 153, 171–72

Progressive era, xv, 70, 153

Public Health Department, Colorado, 66

public history, xv, 3, 35, 44, 95, 90, 95, 103, 107, 109, 171–72

Public Utility Regulatory Policy Act (PURPA), 131, 177

Pullman railroad cars, 61

Purgatory Ski Resort, 62

Puritans, xv, 12

Pursell, Carroll, 52

Putin, Vladamir, 104

Quinnipiac (People), 36

Quinnipiac River, 36

Radiation Exposure Compensation Act (1990), 106

Radkau, Joaquim, 12

railroad train, 2, 14. *See also* heritage railroads

Rainbow Bridge, Utah, 33

Raponi, Richard, 144

Reading, Pennsylvania, 20

Recreational Demonstration Area (RDA), 20

Reddy Kilowatt, 153

refining, 7, 14, 79, 97

Reframing History. *See* AASLH

Reid, Debra, xvi

renewable energy:
geothermal, 129, 132, 176;
histories of, 13–133;
wind, 35, 39, 87, 88, 129–30, 135–38, 140–42;
solar, 79, 83, 86, 88, 129–30, 138–40, 142–45;
interpreting, 133, 135

Rensselaer County Historical Society, Troy, New York, 27

Rentzhog, Sten, 6

Richmond, Linda, 111–12

Righter, Robert, 129, 131–32

rigs (oil), 88, 129

Rilling, Donna, 35

Rios, Alberto, 45

River Transformed (exhibit), 41

Roadside America, 95–96

Rockefeller, John D., 82, 92, 175

Rockwell, Norman, 47

Rogers, Tom, 132

Roosevelt Dam (now Theodore Roosevelt Dam), 158, 160

Rose, Julia, xvi

Rosen, William, 52

Rural Electrification Administration (REA), 131, 155, 157, *163*, 176

Russia, xiv, 83, 104, 141

Rutkow, Eric, 12

Sabin, Paul, 83

Sabino (The), 53–60, *54*, *58*, *59*, 71, 72

Salazar, Epimenio, 163

Salazar, Rosalia, 162

Salisbury Iron District, 14–15, 18–19

Salt River Project (SRP), xii, 43–46, *45*, 158, *159*;
Arizona Canal, 44;
Arizona Falls, *43*, 44–45

Salt River Valley, 43–45

Sample, Paul Starrett, 46;
Norris Dam, 46

San Juan National Forest fire, 66

Savery, Thomas, 81

Saylor, John, 154

About the Author

Leah S. Glaser, PhD, is a professor of history and coordinator of the Public History Program at Central Connecticut State University. She earned her BA from Tufts University and her MA and PhD in history and public history from Arizona State University. She has worked in the field of public history and historic preservation for the last thirty years, consulting with municipal, state, and federal agencies, including positions at the United States Bureau of Reclamation and the National Park Service. For the last twenty years, while publishing around waterpower and electricity, she has closely worked with the National Council on Public History to emphasize issues of environmental sustainability through conferences, committee work, and publications.